The Latin American Studies Book Series

The Latin American Studies Book Series promotes quality scientific research focusing on Latin American countries. The series accepts disciplinary and interdisciplinary titles related to geographical, environmental, cultural, economic, political, urban and health-related research dedicated to Latin America. The series publishes comprehensive monographs, edited volumes and textbooks refereed by a region or country expert specialized in Latin American studies.

The series aims to raise the profile of Latin American studies, showcasing important works developed focusing on the region. It is aimed at researchers, students, and everyone interested in Latin American topics.

Submit a proposal: Proposals for the series will be considered by the Series Advisory Board. A book proposal form can be obtained from the Publisher, Andrea Brody-Barre (andrea.brody-barre@springer.com).

Atilio Francisco J. Zangrando · Angélica M. Tivoli ·
María Paz Martinoli · Daniela V. Alunni
Editors

Archaeology of the South Coast of Tierra del Fuego

Cultural Niches and the Diversity of Hunter-Gatherer-Fishers

Springer

Editors
Atilio Francisco J. Zangrando
CADIC-CONICET
Ushuaia, TDFeIAS, Argentina

Angélica M. Tivoli
CADIC-CONICET
Ushuaia, TDFeIAS, Argentina

María Paz Martinoli
CADIC-CONICET
Ushuaia, TDFeIAS, Argentina

Daniela V. Alunni
CADIC-CONICET
Ushuaia, TDFeIAS, Argentina

ISSN 2366-3421 ISSN 2366-343X (electronic)
The Latin American Studies Book Series
ISBN 978-3-032-09486-5 ISBN 978-3-032-09487-2 (eBook)
https://doi.org/10.1007/978-3-032-09487-2

This Springer imprint is published by the registered company Springer Nature Switzerland AG
The registered company address is: Gewerbestrasse 11, 6330 Cham, Switzerland

Preface

Tierra del Fuego, remote and rugged, has long fascinated scholars of human prehistory. This landscape—shaped by the interplay of sea and mountains, abundance and scarcity—offers a powerful natural laboratory for understanding the adaptive strategies of foraging populations across time. In *Archaeology of the South Coast of Tierra del Fuego: Cultural Niches and the Diversity of Hunter-Gatherer-Fishers*, the authors present an interdisciplinary work that significantly deepens our knowledge of human-environment interactions at the southern tip of the inhabited world.

As editors, we are pleased to present a book that not only integrates diverse lines of archaeological evidence, but also engages deeply with issues of cultural variability, environmental adaptation, and human resilience. The southern coast of Tierra del Fuego—bounded by the Beagle Channel, battered by sub-Antarctic winds, and shaped by millennia of human activity—offers a unique context for exploring the long-term dynamics of hunter-gatherer-fisher societies. The chapters in this book reflect a multi-scalar and interdisciplinary approach to the archaeological record of the southern Fuegian coast. Drawing upon over fifteen years of sustained research, the authors examine long-term demographic trends, changing subsistence strategies, technology, and patterns of mobility and social interaction. The topics addressed include zooarchaeological analyses of faunal use, isotopic studies of human diets and food webs, diachronic studies of shellfish exploitation, lithic sourcing, and the spatial-temporal distribution of archaeological sites. This integrative approach provides a comprehensive understanding of how hunter-gatherer-fisher populations responded to environmental variability and social challenges.

A key theme that runs through the volume is the concept of cultural niche diversity. Rather than viewing Fuegian foragers as static or homogenous, the studies in this book emphasize their ecological plasticity, social dynamism, and regional variation. This perspective challenges long-standing ethnographic generalizations and contributes to a more flexible understanding of cultural evolution in foraging contexts. Moreover, the integration of theoretical frameworks such as niche construction theory and historical ecology positions this volume at the intersection of current debates in archaeology and the environmental humanities.

As editors, we hope that this book will serve not only as a synthesis of important regional research but also as a foundation for future work in southern Tierra del Fuego and beyond. We believe it will be of interest to archaeologists, anthropologists, historical ecologists, and scholars working on coastal and high-latitude adaptations worldwide. Above all, we hope it underscores the enduring relevance of Tierra del Fuego as a landscape of archaeological significance and as a case study in the resilience and creativity of human lifeways.

Ushuaia, Argentina

Atilio Francisco J. Zangrando
Angélica M. Tivoli
María Paz Martinoli
Daniela V. Alunni

Acknowledgements

The research presented in this volume reflects the collective effort of many individuals and institutions over the course of several years. We are deeply grateful to all those who contributed their time, knowledge, and support to the development of this work.

We are grateful to the research institutions and funding agencies that have supported the fieldwork and analytical studies on which this volume is based. In particular, we acknowledge the Centro Austral de Investigaciones Científicas (CADIC-CONICET) for its long-standing institutional support. We also extend our thanks to the Institute of Geochronology and Isotopic Geology (INGEIS-UBA-CONICET) for their support in stable isotope research and for hosting doctoral students. Consejo Nacional de Investigaciones Científicas y Técnicas (CONICET), the Agencia Nacional de Promoción de la Investigación, el Desarrollo Tecnológico y la Innovación, "Marine Ventures" Project (Research Council of Norway—no 208828—Dir. Hein Bjerck), and other national and international programs provided essential funding for archaeological and interdisciplinary research in Tierra del Fuego.

Special thanks go to the colleagues, technicians, and students who assisted in field excavations, laboratory analyses, and logistical coordination—often under challenging conditions in remote areas of the Fuegian coast. Their behind-the-scenes efforts made much of this work possible. We especially thank our teammates—Ana Butto, Mercedes Corbat, María del Carmen Fernández Ropero, Jonathan Nye, and Germán Pinto Vargas—and our colleagues from the Marine Ventures project—Hein B. Bjerck, Heidi M. Breivik, Jo Sindre P. Eidshaug, Joan Negre, and Ole Risbøl—for their continuous support and collaborative efforts over the past ten years. We also extend our gratitude to our colleagues from the Laboratorio de Geomorfología y Cuaternario, CADIC—Marilén Fernández, Ramiro López, and Alejandro Montes.

Furthermore, we would like to express our sincere appreciation to the researchers and collaborators who have contributed to the interdisciplinary studies presented in this volume. Their insights and dedication have greatly enriched the quality of this work: Ricardo Bastida, Florencia Borella, Karen Borrazzo, Luis Borrero, Laura Caruso Fermé, Marianne Christensen, Cristian Crespo, Hernán De Angelis, Cristina Dejean, Marilyn Fogel, María Amelia Gutiérrez, Dominique Legoupil, Giorgio

Manzi, Seth Newsome, Héctor Panarello, Viviana Quse, Luciana Riccialdelli, María José Saletta, Camilla Speller, Mary Ann Tafuri, and Andrew Ugan.

We are deeply grateful to archaeologists Luis Abel Orquera and Ernesto Piana, whose pioneering research and enduring commitment to the archaeology of Tierra del Fuego have provided a solid foundation for the future investigations and analytical approaches developed in this volume.

The Yagan Paiakoala Indigenous Community and local residents of Tierra del Fuego have shown generosity and interest in the research. We also appreciate the continued support provided over the years by the authorities of the Museo del Fin del Mundo, and especially by Lucas Sosa, who is responsible for the archaeological collections. The assistance of Lucas Turnes, from the Dirección General de Patrimonio y Museo, Provincia de Tierra del Fuego (Argentina), was likewise essential.

Finally, we thank the anonymous reviewer of this book for the valuable suggestions, which have allowed us to improve its content. This book is the result of collaboration, dialogue, and shared curiosity about the human past of Tierra del Fuego. We hope it contributes meaningfully to the broader understanding of human adaptability, ecological relationships, and cultural diversity in coastal and high-latitude settings.

Contents

Contributors

Daniela V. Alunni CADIC-CONICET, Ushuaia, TDFeIAS, Argentina

Dánae Fiore FFyL-UBA, CABA, Argentina;
AIA-CONICET, CABA, Argentina

Anna Franch Bach CADIC-CONICET, Ushuaia, TDFeIAS, Argentina;
UAB, Barcelona, Spain

Oriana Hernandez Herrero CADIC-CONICET, Ushuaia, TDFeIAS, Argentina

Sayuri Kochi Department of Pre- and Protohistory, Institute of Ancient Studies, Johannes Gutenberg University Mainz, Mainz, Germany;
Instituto de Geocronología y Geología Isotópica (CONICET/UBA), CABA, Argentina

María Paz Martinoli CADIC-CONICET, Ushuaia, TDFeIAS, Argentina

Suray A. Pérez FFyL-UBA, CABA, Argentina

Juan F. Ponce CADIC-CONICET, Ushuaia, TDFeIAS, Argentina;
ICPA-UNTDF, Ushuaia, TDFeIAS, Argentina

Augusto Tessone Instituto de Geocronología y Geología Isotópica (CONICET/UBA), CABA, Argentina

Angélica M. Tivoli CADIC-CONICET, Ushuaia, TDFeIAS, Argentina

Martín M. Vázquez CADIC-CONICET, Ushuaia, TDFeIAS, Argentina;
UNTDF, Ushuaia, TDFeIAS, Argentina

Atilio Francisco J. Zangrando CADIC-CONICET, Ushuaia, TDFeIAS, Argentina;
FFyL-UBA, CABA, Argentina

Chapter 1
Modelling Cultural Niches of Foraging Populations Along the South Coast of Tierra del Fuego: An Introduction

Atilio Francisco J. Zangrando, María Paz Martinoli, Daniela V. Alunni, and Angélica M. Tivoli

Abstract This chapter examines the long-term adaptive strategies of hunter-gatherer-fisher populations inhabiting the southern coast of Tierra del Fuego, with a focus on the construction and variability of cultural niches across diverse ecological and social landscapes. By integrating archaeological, ethnographic, ecological, and isotopic data, we aim to understand how Fuegian groups responded to environmental variability, including fluctuations in marine productivity, effects of geographic isolation, and interactions across inland-coastal gradients. This approach underscores the dynamic and context-dependent nature of cultural niches, challenging rigid ethnographic categorizations and highlighting the ecological plasticity of hunter-gatherer-fisher societies in coastal and island environments.

1.1 Introduction

The ethnographic studies of Tierra del Fuego have profoundly influenced archaeological perspectives for decades, particularly concerning the variability in forager behaviours in the region. Although ethnographic knowledge has driven significant theoretical and methodological advancements in coastal archaeology and remains

A. F. J. Zangrando (✉) · M. P. Martinoli · D. V. Alunni · A. M. Tivoli
CADIC-CONICET, Ushuaia, TDFeIAS, Argentina
e-mail: fzangrando@conicet.gov.ar

M. P. Martinoli
e-mail: mpmartinoli@conicet.gov.ar

D. V. Alunni
e-mail: daniela.alunni@conicet.gov.ar

A. M. Tivoli
e-mail: tivoli.angelica@conicet.gov.ar

A. F. J. Zangrando
FFyL, UBA, CABA, Argentina

A. F. J. Zangrando et al. (eds.), *Archaeology of the South Coast of Tierra del Fuego*, The Latin American Studies Book Series, https://doi.org/10.1007/978-3-032-09487-2_1

essential for understanding the recent historical processes of native Fuegian communities, it has also contributed to the entrenchment of rigid cultural and adaptive categories in interpretations of pre-contact periods. As in other coastal regions worldwide (Erlandson 2001; Bailey and Milner 2002–2003), these categories were used to distinguish between hunter-gatherer and fisher (HGF) communities, primarily based on their mobility patterns and intensity of marine resources exploitation. Such categories proved useful in archaeological research for differentiating groups that hunted in open waters using advanced technologies, such as harpoons, from those who primarily used coastal environments without specialized gear for aquatic activities (e.g. Orquera and Piana 1999a: 96). However, a notable consequence of this approach has been the reduction of cultural variability into simplistic dichotomies (e.g. terrestrial-coastal, coastal-littoral) which continue to heavily influence archaeological interpretations today (Zangrando et al. 2014a; Borrero and Martin 2023). This cultural simplification of Tierra del Fuego presents challenges when attempting to analyse the range of behaviours and adaptive strategies exhibited by native populations in their interactions with the diverse ecosystems of the Fuegian region. It fails to capture even the variability documented during the brief period covered by ethnographic records. Actually, it is not surprising given that contacts with European groups may have accelerated different social changes in the native populations (Borrero 2011).

While acknowledging that ethnographic and historical information is pivotal to the archaeology of Tierra del Fuego, this book offers a different perspective to analyse the long-term history of these societies. Our main goal of the archaeological research conducted over the past 15 years in southern Tierra del Fuego (Fig. 1.1) has been to understand the ecological adaptability and capabilities of foraging societies, focusing on the factors driving changes in parameters of cultural niche construction. The development of a cultural niche depends on the interaction of environmental, social, and genetic processes (Laland et al. 2001; Boyd et al. 2011). Therefore, changes can be expected in the short term through specialized and generalized responses to resource availability, depending on ecological fluctuations (Binford 2001: 173), or through long-term processes of specialization and diversification as populations respond to changing natural and social environments (Zangrando 2009a, b). Foragers are expected to display technological innovations when employing generalized strategies. In contrast, under highly specialized conditions, their technological behaviour tends to be conservative, favouring consistent, productive, and well-established methods (Fitzhugh 2001).

Extensive archaeological and ethnographic evidence indicates that the native populations of southern Tierra del Fuego consistently used coastal and marine ecosystems. However, within this broadly generalist framework, local groups also exhibited the capacity to specialize their behaviours within these environments (Alunni et al. 2020; Kochi 2023). This capacity has allowed forager populations in the region, throughout their evolutionary history, to adapt to diverse environmental conditions by adjusting the natural and cultural parameters of their niche configuration (Zangrando 2009a, b: 27, 2018; Tivoli and Zangrando 2011; Vázquez et al. 2011; Nye et al. 2020; Zangrando et al. 2022a, b; Kochi 2023; Martinoli and Zangrando 2024). From this

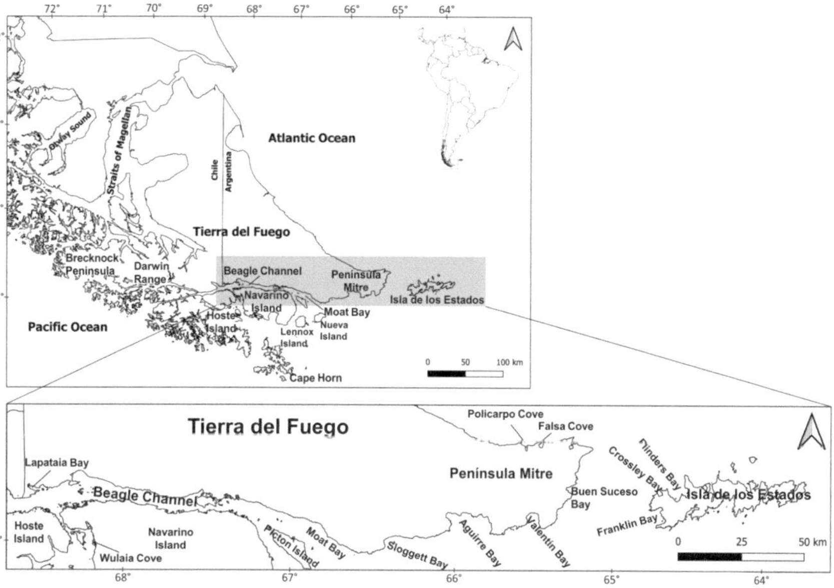

Fig. 1.1 Study area and main geographical references of Tierra del Fuego

perspective, this chapter introduces the regional research problem by reviewing the most relevant information and concepts for understanding the cultural niche construction of HGF populations in southern Tierra del Fuego. First, we present a brief review of ethnographic and historical knowledge that reveals the diversity of geographical, demographic, and social parameters of the forager communities, mainly between the seventeenth and nineteenth centuries. Secondly, we outline the variables for the analysis of the cultural niche of these populations based on archaeological evidence, which constitute the starting points for the analyses and discussions developed in the following chapters.

1.2 Population Dynamics of the HGF According to Ethnographic Record of Tierra del Fuego

Historical records from European expeditions to southern Tierra del Fuego, beginning in the seventeenth century and becoming increasingly frequent and detailed during the eighteenth and nineteenth centuries, document the presence of HGF communities exhibiting distinct physical, technological, and linguistic characteristics. While sorting of these populations varied over time, those identified by the late nineteenth and early twentieth centuries—namely, Yagan (also referred to as Yámana), Haush (or Manekenk), and Selk'nam (or Ona)—remain widely recognized today (Fig. 1.2). All these communities were highly mobile, although they differed in their forms of movement: the former moved by sea using bark canoes, while the others had residential movements by land. However, certain historical records challenge this stereotypical view, portraying a more complex and flexible pattern of mobility among these groups. In the first half of the nineteenth century, Fitz-Roy (2016 [1839]: 136–138) clearly identified the 'canoe people' as Tekeenica, noting that they inhabited the southeastern part of the Fuegian archipelago. He also observed that some families within this group did not reside along the coast but instead relied on guanacos, birds, and berries for subsistence (Fitz-Roy 2016 [1839]: 174). Based on accounts by Orundellico (Jemmy Button), a Yagan informant, Fitz-Roy reported that the Selk'nam made annual raids to the shores of the Beagle Channel, occasionally capturing canoes to travel to Navarino Island (Fitz-Roy 2016 [1839]: 187–188). Other ethnographers, such as Cooper and Furlong (1917), also documented the maritime abilities of Selk'nam groups during hunting expeditions south of the Beagle Channel. Thomas Bridges, who lived for several years on the southern coast of Tierra del Fuego in close contact with native populations, described on two occasions groups that travelled by canoe during the summer, and shifted to overland movements in winter, reaching the eastern tip of Tierra del Fuego (SAMM 1874: 190; 1881: 217). Bridges (SAMM 1886: 217) further noted that the coastline between Moat and Buen Suceso bays was frequented by Selk'nam groups, although these may have been Haush populations (Orquera and Piana 1999b: 82).

Ethnographic data remain inconclusive regarding the geographic distribution of these communities in southeastern Tierra del Fuego (Fig. 1.3), particularly along the coastal sector between Moat and Valentín bays. Numerous European colonists suggest a predominant presence of terrestrial groups east of Sloggett Bay. Willis (SAMM 1876: 223) reported the absence of canoes east of Sloggett, while Martial (2007 [1888]) documented the presence of Selk'nam groups in the same region. Similarly, Spegazzini (1882: 160) noted that Yagan territorial occupation did not extend beyond Sloggett Bay. Furlong (1917) described Haush territory as ranging from Sloggett Bay to Policarpo Cove, later extending it northward to Cape San Pablo. Thomas Bridges estimated Haush territory as spanning from Policarpo Cove to Moat Bay. In contrast, Lefrant (Segura 1973) observed native groups with canoes in Buen Suceso Bay. Hyades and Deniker (2007 [1891]: 425) published a map illustrating the range of canoe groups extending as far as Valentín Bay. Lothrop (1928) defined Haush territory within Peninsula Mitre, between Sloggett Bay and Policarpo Cove, while the Yagan occupation did not extend beyond Sloggett Bay. According to Lothrop (1928), the Selk'nam groups occupied the area south of the Fuegian Andes up to the northern coast of the Beagle Channel. Gusinde described the Yagan territory as forming a 'triangle between Sloggett Bay, the western foothills of the Brecknock Peninsula, and Cape Horn', yet ultimately noted 'depending on where the boundary is drawn', referencing Sloggett or Valentin bays (Gusinde 1986 [1937]). Other authors do not include the southern coast of the Peninsula Mitre within Haush territory (Segers 1891; Coiazzi 1914). Several records indicate that interactions between canoe and land groups were active and dynamic, involving both material exchanges and interethnic marriages (Orquera and Piana 1999b: 512–516; Vidal 2011 [1985]).

Fig. 1.2 Foragers of Tierra del Fuego in post-contact times: **a** Yagans fishing in a macroalgae environment (photograph taken by the Cape Horn Scientific Mission 1882–1883 (Hyades and Deniker 2007 [1891])); **b** Selk'nam group moving along the coasts of Tierra del Fuego (Furlong 1917)

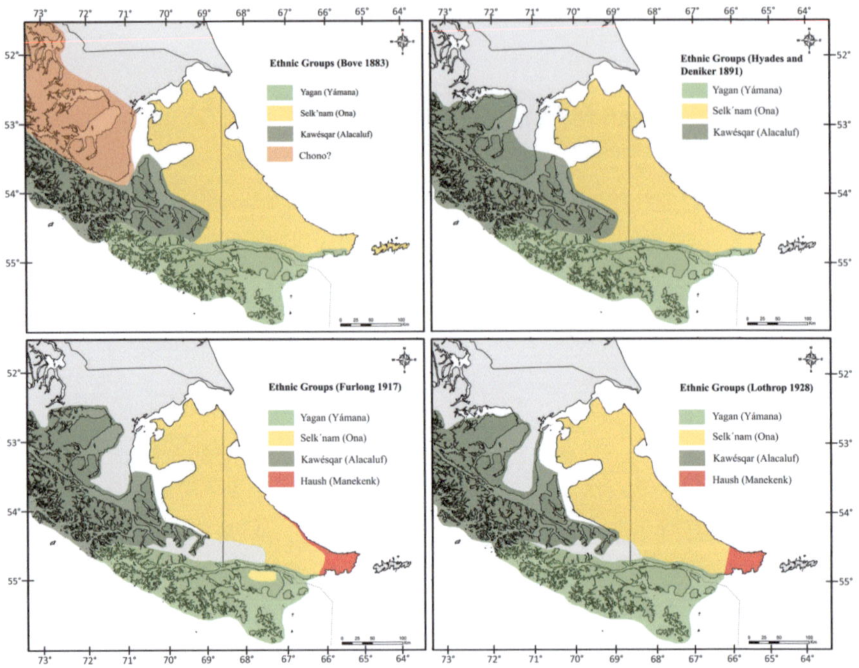

Fig. 1.3 Geographic distribution of native communities in Tierra del Fuego according to ethnographers from the late nineteenth to early twentieth centuries (Bove 1883; Hyades and Deniker 2007 [1891]; Furlong 1917; Lothrop 1928)

Regarding population estimates, Lothrop (1928: 97) calculated that Peninsula Mitre was home to approximately 300 inhabitants prior to European arrival. Furlong (1917: 175) suggested that this number was between 200 and 300 by 1834, but had declined to roughly 100 by 1891. Lucas Bridges estimated the Haush population in the 1880s at no more than 60 individuals. Several authors note that, by the early twentieth century, very few members of this ethnic group remained (Vidal 2011 [1985]: 89–90). In the archipelago south of Tierra del Fuego, T. Bridges (SAMM 1869: 113) reported that approximately 2500–3000 people spoke the Yagan language during the second half of the nineteenth century (SAMM 1880: 74). Gusinde (1986 [1937]) supported Lawrence's estimate that the maximum Yagan population would have been 2500, while Martial (2007 [1888]: 215) estimated a total population of 1300–1500. In September 1883, T. Bridges reported to Hyades (1884: 552) that the Yagan population numbered around 2000, though this figure had been heavily affected by a two-year tuberculosis epidemic. A census conducted by T. Bridges in June 1884 found approximately 1000 survivors (SAMM 1884: 223).

Coastal settlement patterns—whether indicated by group size, dwelling locations, or structural forms—do not offer a definitive basis for distinguishing between the ethnographic groups of southern Tierra del Fuego. References to settlements along the shores of the Península Mitre are limited, though certain historical records provide valuable insights. A notable example comes from the survivors of the *Purísima Concepción* shipwreck, who spent approximately two months interacting with native communities along the northern coast of the peninsula. These survivors reported the presence of huts along the coast near Falsa Cove. During the Spaniards' stay in the cove, they encountered at least three groups comprising 30, 70–80, and 40–50 natives, all camping in wooded areas near the shore. Similar descriptions were provided by Labbé (in De Brosses 1756) and Banks (1896) for the southern coastline of the region, who described groups of 20–50 individuals (Vidal 2011 [1985]: 77). Occasionally, camps were located a few miles inland; for instance, during James Cook expedition in January 1769, the crew found 12–14 huts in a forested area about two miles from the coast at Buen Suceso Bay (Fig. 1.4). Both dome-shaped and conical huts made from logs and covered with thin branches, moss, and peat are described and illustrated for the Península Mitre. Most coastal resource-processing activities occurred in these camps, where mollusc shells were found among the dwellings. As European contact mostly took place during the summer, further insights into seasonal settlement patterns in the Península Mitre remain limited (Vidal 2011 [1985]).

Canoe groups occupied coastal areas year-round, exhibiting a high degree of mobility with seasonal variations in the frequency and range of their movements. Many ethnographers attributed this high residential mobility to local resource depletion, particularly mollusks (Orquera and Piana 1999b). However, Bridges (VSA 1866: 207) observed that in winter, native groups tended to lead a relatively less mobile lifestyle. Similarly, Gusinde (1986 [1937]: 600–601) noted that the Yagans

Fig. 1.4 Native settlement inland Buen Suceso bay, Península Mitre (Hawkesworth 1773)

exhibited greater mobility in summer, when resource diversity was higher. Fitz-Roy (2016 [1839]: 167) and Martial (2007 [1888]: 196) indicated that canoe groups extended their movements to the outer islands during the summer, while seeking shelter in the inner channels during winter.

Detailed descriptions of their dwellings highlight two predominant types: dome-shaped and conical, both constructed with logs, branches, and grasses. Each dwelling typically had a circular or semi-circular base, with diameters ranging from 2 to 5 m and heights around 2 m. These dwelling types were not uniformly distributed across the archipelago south of Tierra del Fuego. Ethnographers and informants often observed that dome-shaped structures were more common in the western areas of the archipelago, while conical structures prevailed in the eastern areas (Orquera and Piana 1999b). The Yagans generally lived in small, isolated family units (Fig. 1.5). Ross (1847: 304) never observed more than fifteen people together, although Snow (1857 I: 341; 1861: 263; 1864: 164) reported a group of 22 people living communally. Similarly, Gardiner (1857) described groups ranging from 13 to over 25 individuals.

Gusinde (1986 [1937]: 354) argued that settlements did not form villages, yet several chroniclers described clusters of dwellings and groups of people. Gardiner (VP 1858: 222) recorded a group of 35, while Weddell (2006 [1825]: 169) encountered 80 people gathered together. At Wulaia Cove, larger aggregations of 100 to 200 people were recorded (Orquera and Piana 1999b: 90). Martial (2007 [1888]: 207) also mentioned groups of around 100 natives. These larger aggregations, however, were not permanent settlements; rather, they formed in response to specific circumstances,

Fig. 1.5 Yagan family in a hut (Hyades and Deniker 2007 [1891])

such as the presence of European expeditions, whale stranding, or ceremonial events (Orquera and Piana 1999b).

Ethnographic information also indicates that the HGF communities of southern Tierra del Fuego exploited a similar variety of marine and coastal resources, both in terms of the capture of pinnipeds, seabirds and fish, and the harvesting of molluscs. The technology employed for resource procurement and processing was also similar among the different societies.

Within the brief period covered by the ethnographic record, it is difficult to identify a clear pattern in the spatial organization of native populations in southern Tierra del Fuego. Additionally, the record reveals ethnic groups with mixed mobility and spatial settlement patterns. It is also important to consider that populations with aquatic mobility engaged in wide-ranging movements at certain times of the year or as part of specific logistical strategies. In sum, the cultural landscape observed in southern Tierra del Fuego during the eighteenth and nineteenth centuries was notably dynamic. Consequently, using categories such as 'marine' or 'canoe' and 'coastal' or 'terrestrial' risks oversimplifying these complex dynamics, overlooking the interactions between different lifestyles.

1.3 The Bypass to Archaeology

Studying the extent of a cultural niche leads to identifying and interpreting changes in human behaviour at varying spatial and temporal scales, which have been the subject of numerous discussions over the last decade. The range of species consumed, the uniformity of prey exploited over time, the range of trophic levels represented in prey and feeding areas are some of the variables that can be analysed in the archaeological record. For example, faunal evidence provides a rich source for investigating these long-term processes (Broughton et al. 2010). There are various analytical procedures by which it is possible to analyse this evidence (Hill 1973; Byrd 1997) and to explore the nature of the generalization (or specialization) of the archaeological record (Grayson 1984; Broughton 1994; Betts and Friesen 2004). This involves contemplating its nature and the problems associated with sample size, excavation techniques, biases in bone preservation and other taphonomic issues that can significantly influence our estimates of diversity (Cannon 1999, 2001; Lyman 2003).

Such discussions at long-term scales are fruitful in archaeology, but the construction of a cultural niche is also affected by processes operating at the level of the individual, local groups or bands; these processes inherently reflect different dimensions of social organization (Butzer 1982; Bailey 1983). For instance, a pertinent question for coastal environments—where archaeological evidence typically indicates a wide range of resource exploitation—is whether the HGF population consisted of generalists who consumed a broad spectrum of available foods, or individuals who each consumed different types of foods (Bearhop et al. 2004). Distinguishing between generalizing forms of how HGF communities interacted with marine ecosystems in the past is a key issue, as the ability to generalize and specialize simultaneously may have competitive implications under fluctuating environmental and ecological conditions (Roberts and Stewart 2018). The comparison between information at population and individual scales, where information on richness and compositional evenness of archaeological assemblages can be combined with dietary information on individuals, provides a useful and robust parameter for analysing the extent of the human niche in the past. Stable isotopes from $\delta^{13}C$ and $\delta^{15}N$ measured on collagen allow us to establish the type of protein intake averaged over the last years in the life of an individual. In contrast, food remains in archaeological deposits often refer to time-averaged records of food consumption by individuals over centuries or even millennia. This difference in resolution can, of course, be exploited to contrast dietary variation at the individual level with the consumption patterns of a population for a given period (Bailey and Milner 2002–2003; Barberena et al. 2004; Kochi 2023).

Another example can be seen through the exploitation of molluscs and the formation of shell middens. The accumulation of these deposits, owing to their temporal structures, occurs at a variety of tempos and involves behavioural, formative and evolutionary processes, among others. Archaeological excavations in the Beagle Channel (Tierra del Fuego, Argentina) have shown that the formation of shell middens resulted from repeated occupations (Orquera et al. 1977; Orquera and Piana 1986–1987, 1996, 1999a; Yesner 1990; Estévez and Vila 1995, 2006, 2007; Piana et al. 2004; Álvarez et al. 2009; Zangrando 2010; Zangrando et al. 2016). These reoccupations are indicated both by short term but continuously repeated shell accumulations and by reuse of the same site after long-term pauses on century or millennial scales. As such, the formation of shell middens has been seen as the result of long-term practices in which the accumulation of shell deposits promoted their subsequent reuse (Mackie 2001; Piana and Orquera 2010). Shell middens occur as dense agglomerations widely distributed along the shores of the Beagle Channel, to such an extent that they represent a prominent anthropogenic feature of the coastal environment. These dense shell deposits functioned as distinct nodes in the mobility patterns of HGF societies.

Mollusc exploitation is a widely debated topic in coastal archaeology, mainly because it involves discussions of paleodemographic conditions and long-term environmental variation (Mannino and Thomas 2002). While predictions of overexploitation are fulfilled in some cases (e.g. Erlandson et al. 2008; Klein and Steele 2013), studies do not usually focus on identifying other spatio-temporal scales at which an impact on the mollusc stock can be archaeologically identified (although see Parkington et al. 2013). It has been observed, however, that a progressive decrease in the modal size of mussel shells deposited in the different occupation events that make up a deposit may be reflecting other processes taking place at a smaller scale and may be structured by social practices in the form of harvesting (Jones and Richman 1995). The impact on mollusc populations can also occur in the short term, spanning weeks or months, as a result of regular or seasonal movements between different coastal areas within a region, particularly when there is a reduction in the availability of specimens of optimal size (Orquera and Piana 2001). This may also coincide with shifts in taxonomic composition, favouring a higher proportion of smaller Mytilidae species (Pérez 2020). Thus, shell midden formation as cumulative palimpsests (Bailey 2007) represents a dynamic interaction between top-down factors, such as demographic and environmental influences, and bottom-up factors shaped by diverse social practices, including gathering techniques. Understanding how these factors come together in the human exploitation of molluscs and in the cumulative discarding of shells allows us to know the modification of coastal environments as part of a process of niche construction throughout the Holocene (Zangrando 2018; Zangrando et al. 2021).

1.4 Cultural Niche Model

The construction of a cultural niche requires consideration of two forms of environmental variability (Winterhalder and Smith 1992: 8–9). Strategic contexts or variables refer to environmental constraints whose value is conditioned by systemically interconnected responses and counter-responses. The social environment is often recognized as strategic in its functioning because the adaptive value of behaviour depends on the actions and reactions of others. The studies outlined previously allow us to analyse changes in these parameters through archaeological evidence. Parametric variability refers to the changing nature of physical conditions that impose selective constraints and opportunities for survival and reproduction. These variables are independent of any feedback from behavioural responses and are usually modelled probabilistically. In the context of our case study, such variables are expressed through biogeographical conditions. The corridor comprising the southern coast of Tierra del Fuego, from Lapataia Bay (central Beagle Channel) to Isla de los Estados, offers a margin of macro-regional variation that allows a series of expectations for parametric analysis (Fig. 1.1). Within this corridor, two key biogeographical features structure distinct territorial, ecological, and social mechanisms. On the one hand, the projection of the Fuegian Andes runs parallel to the whole of this sector, with elevations decreasing in a west–east direction. On the other hand, with the same geographical orientation, there is less and more exposure to marine conditions in a gradient that runs between inland channels (Beagle Channel), coasts with greater exposure to sea fronts (Península Mitre) and offshore environments (Isla de los Estados). Five factors involved in the construction of the cultural niche within this environmental variability are developed and substantiated below.

1.4.1 Population Distribution Geometry

Different marine and coastal environments influence the interrelationship of human population nodes in various ways, where isolation and interaction are presented as a continuum in the exchange of knowledge and materials across space (Boomert and Bright 2007; Broodbank 2008; Erlandson 2008). Within this dynamic, the geometry of human population distribution is an important factor for survival opportunities at the local or regional level. Demographic and cultural continuity is expected to be greater in the central sectors of such distributions with higher population densities and a geographic matrix that provides greater interaction between nodes of occupation, such as an archipelago environment (Fitzhugh 1997). These situations favour gene flow and the transmission of information and thus promote the accumulation of knowledge across generations (Boyd et al. 2011; Hill et al. 2014). These advantages are more restricted in linear or peninsular environments as fewer nodes interact. These configurations are expected to be more exposed to segmentation and genetic isolation conditions (Fitzhugh 1997).

Variations in population density in a given space can result from long-scale spatial reorganizations and redistributions, even without significant changes in the overall size of a population (Fitzhugh et al. 2020). Based on available chronological information, Legoupil and Fontugne (1997) observed that the distribution of occupations was not spatially continuous during the last 7000 years in the Tierra del Fuego archipelago. These authors pointed out that the occupation of this space would have occurred in two stages, differentiating between the existence of core populations in the interior of the channels, and a later concentric expansion towards more peripheral sectors of the archipelago. Within this model, the Beagle Channel is presented as an ecotone where the exploitation of terrestrial and maritime resources was equally possible and where the most extended occupation sequences have been recorded. Scheinsohn (1997) also argued that southern Tierra del Fuego would have been populated ('saturated') more rapidly than other sectors of the archipelago. The subsequent dispersal movement would be evidenced by the presence of occupations with radiocarbon ages after 2800 cal yr BP and located in the chain of islands that form the periphery of the archipelago: Cape Horn and Isla de los Estados.

Demographic conditions among HGF societies can vary markedly by location. Even when faced with the same environmental and social changes on a global scale, the population may decline in some areas but grow in others. For example, the geometry of population distribution on the southern coast of Tierra del Fuego could vary according to the zones of contact between native and European groups and change rapidly over time. At the beginning of this process, contacts are mostly on the coasts more exposed to the oceans, while the interaction between both groups is more stable later in the nineteenth century within the interior channels when Europeans had already practically explored the region and the first permanent settlements were established (Borrero and Martin 2023). To understand the long-term history of HGF communities in marine environments, it is not only the increase in population size that matters, but also the spatial organization of occupations and the processes of formation of archaeological deposits that also play a role (Bailey 2007). These aspects of population dynamics were analysed on the basis of archaeological (Zangrando et al. 2009), genetic (Crespo et al. 2020) and stable isotope evidence (Kochi 2023) and their implications are considered in the following chapters.

1.4.2 Coast-Inland Interaction

Both terrestrial and marine ecosystems are equally dominant in the region. Although in the south of Tierra del Fuego they present important contrasts in terms of the diversity of animal and plant species and in the extensions of their food chains, both environments were crucial for human adaptation. Marine and coastal ecosystems offer greater faunal abundance, while the interior contains a lower animal biodiversity, where the guanaco (*Lama guanicoe*) is the only relevant resource for human subsistence (Orquera and Piana 1999a). Inland water bodies, such as rivers and lakes, have few native species of fish and none of molluscs. The Magellanic forest extends to the coastline and provides, in addition to firewood, key raw materials for the construction of dwellings, of boats and technological production such as hunting weapons, containers, or ritual implements.

We know very little about the forms of coast-inland interaction. The Fuegian Andes would have presented a highly variable scenario with dissimilar possibilities and conditions for human circulation along the southern margin of Tierra del Fuego. In the western and central sections of the Beagle Channel, within an archipelagic landscape, the highest elevations are recorded, featuring narrow valleys often encroached upon by glaciers from the expansive Darwin Range icefields. In contrast, conditions from Moat Bay to the eastern end of the Beagle Channel reveal a significantly lower mountain backdrop, interrupted by broad valleys that run transversely to the coastline. These valleys may have served as mobility routes, potentially linking coastal and inland occupation sites. The inland region of southeastern Tierra del Fuego is distinguished by extensive peatland fields. Drawing on information from Lista (1887), Vidal (2011 [1985]: 47) suggests the possible existence of passes used by native groups through valleys linking the heads of Buen Suceso, Valentín, and Aguirre bays. These routes were later utilized by explorers from Península Mitre (Payró 1898; Chapman and Hester 1973). Another example of mobility is seen in the creation of 'porteos' (Borrero et al. 2020). These terrestrial routes were established to overcome land barriers within marine circuits and required terrain preparation to transport canoes across them (Prieto et al. 2000; Martial (2007 [1888]); Maximiano Castillejo 2016). The clearing and maintenance of 'porteos' reflects a need for safe and time-efficient passage, embodying a practical form of niche construction.

Another possibility to be considered along the southern margin of Tierra del Fuego is whether inland settlement occurred in an integrated mode with coastal environments, rather than always accepting a coastal pattern of occupation with logistical inland uses. The possible presence of settlements at significant distances from the coast has not yet been systematically explored, although this situation has been recorded at historic times (Vidal 2011 [1985]).

The use of inland spaces may have been linked to the provisioning of various resources. The exploitation of guanacos recorded in several coastal contexts implies that inland logistical forays were undertaken to hunt this prey (Alunni et al. 2020). Abundant ethnographic information indicates that hunting groups could stay several days in the interior of the valleys travelling long distances (~ 25 km) from the coast to capture this resource (Orquera and Piana 1999a: 141–142). The acquisition of raw material for the production of technology could be another factor. The site Altos del Varela I, at the top of Cerro Petersen (600–700 m), is located 10 km in a straight line from the northern shores of the Beagle Channel, and it evidences the processing of rocks corresponding to the Lemaire Fm. (Álvarez et al. 2010). Two sites have also been identified in the interior of Navarino Island, the quarry/workshop sites Cerro Bandera 1 and Paso Laguna del Salto 2, both more than 6 km from the coast in a straight line (Borrero et al. 2020). These latter cases indicate that this form of inland resource use does not only respond to terrestrial mobility circuits, as it might imply the case north of the Beagle Channel. Obtaining bark, a fundamental raw material for the production of canoes during the nineteenth century, may have involved journeys of up to 8–10 km inland (Gusinde 1986 [1937]: 424; Hyades and Deniker 2007 [1891]: 159).

1.4.3 Isolation

The southern coast of Tierra del Fuego, including the Beagle Channel, Peninsula Mitre and Isla de los Estados, is a bioceanic corridor that played a key role in the transmission of knowledge and materials throughout the Holocene (Zangrando et al. 2009). However, this dynamic cannot be considered geographically or temporally uniform in the diversity of conditions projected in the region: the configuration of this environment presents interior channels, such as protected contexts for mobility and resource provisioning, to offshore environments with greater constraints and uncertainties for human activity. In the case of southern Tierra del Fuego, unlike other islands, Isla de los Estados presents an offshore condition and it was sporadically occupied throughout the Holocene (Chapman 1987; Horwitz 1990, 1993; Horwitz and Weissel 2011). Isla de los Estados is an island far from the main land mass (in this case Isla Grande de Tierra del Fuego) whose occupation implied the crossing of a 30 km wide channel that was difficult to navigate due to the encounter of marine currents. The isolation of any island or sector of the archipelago (as in other contexts of human society) is a relative concept. Local groups or individuals inhabiting island spaces have the potential to be more or less isolated, and the degree to which they interact or isolate themselves is affected by multiple environmental and cultural factors operating at various scales (Broodbank 2008; Erlandson 2008; Fitzpatrick and Anderson 2008). Investigating whether the sporadic occupation of Isla de los Estados was influenced by isolation due to physical or social barriers, and whether the occupying groups remained isolated or engaged with other groups during this period, is crucial for understanding the historical development and niche diversity of the regional populations (Zangrando et al. 2024).

Isla de los Estados is one of the few areas in the archipelago that was uninhabited by native groups at the time of European arrival. This is particularly noteworthy given that it covers over 500 km^2 and is rich in biodiversity, particularly in marine species, largely due to the presence of bird and pinniped colonies. Archaeological evidence suggests that Isla de los Estados was occupied for a brief period between 2800 and 1375 cal yr BP. Archaeological sites have been identified in three bays—Franklin, Crossley, and Flinders—located at the western end of the island. It has been proposed that the use of offshore spaces like Isla de los Estados was associated with the seasonal exploitation of seabird colonies (Lefèvre 1993–1994; Legoupil 1993–1994, 1995). This seasonal use for resource provisioning would imply interactions with other inland sectors of the archipelago, where population nodes were likely situated. Based on the available archaeological data, Horwitz (1990: 226) identified two stages of occupation for Isla de los Estados. In the first stage, prior to 1375 cal yr BP, the island was likely integrated into the occupation area and mobility circuits of canoe populations, possibly linked to groups from the eastern Beagle Channel and the islands of Lennox, Picton, and Nueva. The second stage is characterized by the abandonment of the island after that date, likely due to the displacement of marine hunter-gatherers resulting from increased occupation by other hunter-gatherer groups on the Peninsula Mitre. In this context, Isla de los Estados would have become part of a different system, which, during ethnographic times, was expressed through the mythological worldviews of both the Selk'nam and Yagan (Chapman 1987; Horwitz and Weissel 2011).

1.4.4 Paleoecology and Resource Fluctuations

Resource fluctuations appear as active forces that affect the chances of survival, resilience and adaptation in a cultural niche. Changes in the behaviour of animal species, in faunal diversity and in the breadth of food webs usually respond to processes induced by environmental and anthropogenic factors. Consequently, to understand human decisions and their consequences in the past, we need to characterize palaeoecological environments to support our interpretations and models of past foraging behaviour (Bird and O'Connell 2006). The need to investigate the population structure of animal species in the past—particularly before the overexploitation of otariid populations in southern South America during the eighteenth and nineteenth centuries (Bonner 1981; Schiavini 1993; Orquera 2002) and the commercial overexploitation of fishery resources in the South Atlantic (Esteves et al. 2000)—has driven research into the historical ecology of various organisms in the Fuegian archipelago throughout the Holocene.

One of the most relevant discussions for the region has focused on the human-otariid relationship. Two species of pinnipeds are the most distinctive in terms of abundance and distribution in the Fuegian archipelago: the South American fur seal (*Arctocephalus australis*) and the Southern sea lion (*Otaria flavescens*). Both species can be found throughout the year, either as dispersed individuals or in permanent or temporary colonies along coasts and on small islands. Most mating and breeding colonies (rookeries) are currently located on coasts and outer islands of the archipelago, away from the inner channels (Schiavini 1993; Falabella et al. 2009; Milano et al. 2020). Based on the current ecology of otariids and the analysis of seasonality of deaths in archaeological specimens, it has been proposed that the HGF communities inhabiting the Beagle Channel did not significantly impact pinniped populations, as these colonies were located outside the primary foraging ranges of these human populations (Schiavini 1993; Orquera and Piana 1999a). However, subsequent studies found that the representation of pinniped remains declined relative to the abundance of other resources in later periods of the archaeological sequence (Zangrando 2009a, b; Tivoli and Zangrando 2011; Chap. 3), alongside a reduction in the body size of these prey throughout the Holocene (Zangrando et al. 2014a; Martinoli 2018). These trends may indicate a decrease in pinniped encounter rates, potentially linked to various ecological factors influencing prey behaviour in response to environmental changes and human activities (Zangrando et al. 2014b). Furthermore, it has been suggested that the distribution of modern colonies—characterized by a heterogeneous arrangement oriented towards the peripheral sectors of the Fuegian archipelago—could be a consequence of the colonization of the archipelago by hunter-gatherer populations (Vidal and Winograd 1986).

The long-term historical ecology of otariids has been explored through multiple isotopic methods, such as the more typical analyses of bulk carbon, nitrogen, and hydrogen in archaeological remains, as well as more advanced methods, including compound-specific amino acid analyses of protein components (Zangrando et al. 2014b; Nye 2019; Nye et al. 2020; Zangrando et al. 2022a, b). From this information, it could be inferred that there were changes in the ecological niches of pinnipeds throughout the Holocene. Although the trophic level of otariids does not change over time, both *A. australis* and *O. flavescens* shifted from coastal habits to feeding further from the coast. Although some variability in the base of the food web is observed, this does not seem to coincide with climatic changes, leading to the conclusion that humans may have had a greater influence on changes in pinniped behaviour. Although these marine mammals show significant plasticity in the face of human predation, as they maintained their trophic level throughout the Holocene, they possibly shifted their feeding grounds towards more pelagic sectors in response to increased human predation pressure more concentrated in littoral environments (Nye 2019; Zangrando et al. 2022a, b).

When pinniped encounter rates declined due to this factor, the HGF communities inhabiting the Beagle Channel underwent a process of niche width expansion. This process involved increased human exploitation of seabirds and fish, as well as an extension into pelagic areas for resource provisioning (Zangrando 2009a, b; Tivoli and Zangrando 2011). The representation of pelagic fish in the archaeological record, encompassing both small forage fish and larger predatory species, varies from 7200 cal yr BP to contact times. While Fuegian sardine (*Sprattus fuegensis*) and Patagonian grenadier (*Macruronus magellanicus*) are prominent in bone assemblages dated between 7200 and 5000 cal yr BP, remains of common or southern hake (*Merluccius* sp.) and snoek (*Thyrsites atun*) are more prevalent from 3000 cal yr BP to the nineteenth century (Zangrando 2009b). The latter species currently inhabit deeper waters in the marine ecosystem of the Tierra del Fuego archipelago (Lloris and Rucabado 1991). This interpretation is supported by paleoecological studies examining the isotopic ecology of prey. Analyses of fish skeletal remains for $\delta^{13}C$

and $\delta^{15}N$ values in collagen from these species reveal that $\delta^{13}C$ values indicate the pelagic nature of *Merluccius* sp. and *T. atun* compared to *M. magellanicus* and coastal fish taxa in prehistoric times (Zangrando et al. 2016). A more recent stable isotope study, which included carbon and nitrogen values across this taxonomic spectrum, concluded that *Merluccius* sp. and *T. atun* shared a very similar isotopic niche during the Late Holocene (Bas et al. 2019).

It has also been suggested that oscillations in sea surface temperature (SST) and marine primary productivity at different stages of the Holocene may have affected the distribution of species such as *Merluccius* sp. and *T. atun* (Bas et al. 2019, 2020). These variations would have had an impact on the opportunities for hunter-gatherer groups that inhabited the coasts exposed to the Atlantic Ocean to access these resources, possibly due to the displacement of fish populations and their spawning grounds to deeper sectors towards the Late Holocene (Bas et al. 2020). This would have led to a lower recurrence of mass strandings, episodes from which groups moving along the coasts could access this type of resource. In contrast, other groups continued to exploit pelagic species through the use of maritime technology and expanded provisioning areas (Bas et al. 2023). This raises the possibility that fluctuations in SST and marine primary productivity may have affected distinctly different HGF groups with different cultural niches.

1.4.5 Resilience

Resilience is the capacity of a system to experience shocks while maintaining essentially the same function, structure, feedback, and thus identity (Walker et al. 2006). In this way, resilience is key to the sustainability of a cultural niche over time (Hudson et al. 2009). In recent years, this concept has been used to study the behaviour of prehistoric populations (including mobility, subsistence and social systems) in response to environmental factors, such as volcanic activity (Grattan and Torrence 2007; Fitzhugh et al. 2016). However, resilience analysis provides insight into the capacity of humans to absorb and resist stresses from internal and external sources, where stability and change are integral and equivalent components (Walker et al. 2006; Bradtmöller et al. 2017). In archaeology, the study of resilience has been approached mainly from subsistence, demographic trends, social organization and the development of technological innovations (Bradtmöller et al. 2017). Both subsistence, based on the analysis of the diversity and abundance of exploited resources, and demographic trends have been frequently addressed (e.g. Stiner and Kuhn 2006; Nelson et al. 2012; Bicho et al. 2017).

Interaction networks in the social organization of human populations are an important factor promoting resilience (Janssen et al. 2006). Although high or low levels of resilience are not intrinsically linked to conditions of insularity, it has been proposed that social networks are important elements in archipelago contexts (Fitzhugh 1997; Hudson et al. 2009). The existence of social networks in the past has been examined based on the distribution of artefacts with known origins, and various scenarios can be considered (Janssen et al. 2006): (1) ecosystems connected by people through information or material flows, (2) ecosystem networks that are disconnected and fragmented by human actions, and (3) artificial ecological networks created by humans (e.g. 'porteos'). Despite being a key element in the social organization of the peoples we study, we know little about the forms and intensity of interaction in Tierra del Fuego. For example, interaction networks may have had important implications for the transmission of infectious diseases, a key factor in the history of native populations with the establishment of Europeans in the region.

To what extent did the environment and history of the Fuegian archipelago influence the ability of hunter-gatherers to develop resilience in response to social and environmental change? Álvarez et al. (2020) analysed the behaviour of hunter-gatherer groups on Península Mitre during the Late Holocene using multiple lines of archaeological evidence. They examined the environmental fluctuations that occurred during this period, including the Medieval Climate Optimum and the Little Ice Age, and their impacts on these societies, primarily regarding changes in the temporal and spatial distribution of resources. The authors concluded that flexible strategies, which involved the exploitation of a diverse range of fauna, versatile technologies, and a mobility organization structured by resource abundance and predictability, were employed over a span of 1500 years. These strategies likely enhanced adaptability and resilience in the face of fluctuating conditions.

It is likely that the subsistence patterns of hunter-gatherers in the Fuegian archipelago, characterized by the exploitation of a wide variety of resources, provided these populations with resilience in the face of industrial hunting of marine mammals by European and North American companies during the eighteenth and nineteenth centuries (Tafuri et al. 2017). Stable isotope studies ($\delta^{13}C$ and $\delta^{15}N$) conducted on the bone collagen of native individuals inhabiting the region during this period indicated a marine-based diet. This suggests that, although post-contact populations may have experienced a significant reduction in pinniped consumption, they likely compensated by consuming other aquatic resources. However, these dietary changes are not detectable at the bulk collagen level isotopically. Thus, the ecological resilience of Fuegian groups is reflected in their pursuit of nutritional alternatives and adaptive dietary practices over several centuries, as evidenced by the isotopic ratios in the bone collagen of individuals (Tafuri et al. 2017).

Butto and Fiore (2020) present an intriguing case regarding the adoption of Western material culture by Yagan society from the late nineteenth to the mid-twentieth century. Through a comprehensive analysis of photographic records, the authors observed that, while the adoption of Western clothing progressed rapidly, Yagan groups continued to uphold certain traditional clothing practices during the contact period, as evidenced by their use of body ornaments (Orquera and Piana 1999b: 297–324). This indicates that the Yagan were active agents in their own cultural transformation, engaging in various processes of negotiation and resilience that facilitated their physical and cultural survival.

1.5 Conclusion

The proposal presented in this chapter models the behavioural diversity of HGF populations in southern Tierra del Fuego under dynamic cultural and environmental conditions that structure particular historical trajectories. It is not possible to argue for the permanence of cultural configurations in the area in the short time frame implied by the formation of the ethnographic record. In any case, this seems to arise more from a construction generated by ethnographers in the twentieth century (Lothrop 1928; Chapman 1986; Gusinde 1986 [1937]), as opposed to what emerges from the narratives provided by naturalists and missionaries during the eighteenth and nineteenth centuries. As we have pointed out, the cultural niche conception allows us to explore the plasticity and ecological capacities of HGF populations through archaeological evidence in the face of variations in the social and natural environment, without the need to preconceive specific adaptations and ways of life for certain sectors and their projection to earlier Holocene times. Moreover, HGF communities considered within the same form of mobility or even part of the same ethnic group may not share the same niche. Our challenge in recent years has involved analysing variations in the breadth of the cultural niche of populations inhabiting an environment ranging from inland channels to offshore islands throughout the Holocene that may have operated at the scale of the individual, site or region. The following chapters examine these dynamics from various lines of evidence, ultimately demonstrating how they are integrated within an interdisciplinary research project.

References

Alunni VD, Tivoli AM, Zangrando AF (2020) Movilidad y explotación de recursos terrestres por cazadores-recolectores costeros de bahía Moat, costa sureste de Tierra del Fuego (Argentina). Lat Am Antiq 31(3):576–594. https://doi.org/10.1017/laq.2020.27

Álvarez M, Zurro D, Briz i Godino I, Madella M, Osterrieth M, Borrelli N (2009) Análisis de los procesos productivos en las sociedades cazadoras-recolectoras-pescadoras de la costa norte del canal Beagle (Argentina): el sitio Lanashuaia. In: Salemme M, Santiago F, Álvarez M, Piana E, Vázquez M, Mansur E (eds) Arqueología de la Patagonia. Una mirada desde el último confín. Editorial Utopias, Ushuaia, p 903–917

Álvarez M, Pal N, Salvatelli L, Briz i Godino I, Zangrando AF, Bazerque A, De Angelis H, Fiore D (2010) Tallando desde lo alto: un sitio de explotación de materiales líticos de los grupos cazadores-recolectores de la Isla Grande de Tierra del Fuego. Magallania 38(1):297–302. https://doi.org/10.4067/S0718-22442010000100019

Álvarez M, Briz i Godino I, Pal N, Bas M, Lacrouts A (2020) Climatic change and human-marine interactions in the uttermost tip of South America in Late Holocene. Quat Int 549: 197–207. https://doi.org/10.1016/j.quaint.2018.06.007

Bailey GN (1983) Concepts of time in quaternary prehistory. Annu Rev Anthropol 12:165–192

Bailey GN (2007) Time perspectives, palimpsests and the archaeology of time. J Anthrop Archaeol 26(2):198–223. https://doi.org/10.1016/j.jaa.2006.08.002

Bailey GN, Milner N (2002–2003) Coastal hunter-gatherers and social evolution: marginal or central? Before Farming 3–4(1):1–15. https://doi.org/10.3828/bfarm.2002.3-4.1

Banks J (1896) Journal of the Right Honourable Sir Joseph Banks during Captain Cook's first voyage in H.M.S. Endeavour in 1768–71 to Tierra del Fuego, Otahite, New Zealand, Australia, the Dutch East Indies, etc. Mcmillan and Co., London

Barberena R, L'Heureux GL, Borrero LA (2004) Expandiendo el alcance de las reconstrucciones de subsistencia. Isótopos estables y conjuntos arqueofaunísticos. In: Civalero MT, Fernández PM, Guráieb AG (eds) Contra viento y marea. Arqueología de Patagonia. INAPL, Sociedad Argentina de Antropología, Buenos Aires, pp 417–433

Bas M, Briz i Godino I, Álvarez M, Vales DG, Crespo EA, Cardona L (2019) Back to the future? Late Holocene marine food web structure in a warm climatic phase as a predictor of trophodynamics in a warmer South-Western Atlantic Ocean. Glob Change Biol 25:404–419. https://doi.org/10.1111/gcb.14523

Bas M, Salemme M, Green EJ, Santiago F, Speller C, Álvarez M, Briz I Godino I, Cardona L (2020) Predicting habitat use by the Argentine hake *Merluccius hubbsi* in a warmer world: inferences from the Middle Holocene. Oecologia 193:461–474. https://doi.org/10.1007/s00442-020-04667-z

Bas M, Salemme M, Santiago F, Briz i Godino B, Álvarez M, Cardona L (2023) Patterns of fish consumption by hunter-fisher-gatherer people from the Atlantic coast of Tierra del Fuego during the Holocene: human-environmental interactions. J Archaeol Sci 152:105755. https://doi.org/10.1016/j.jas.2023.105755

Bearhop S, Adams CE, Waldron S, Fuller RA, MacLeod H (2004) Determining trophic niche width: a novel approach using stable isotope analysis. J Anim Ecol 73(5):1007–1012. https://doi.org/10.1111/j.0021-8790.2004.00861.x

Betts MW, Friesen TM (2004) Quantifying hunter-gatherer intensification: a zooarchaeological case study from Arctic Canada. J Anthrop Archaeol 23:357–384. https://doi.org/10.1016/j.jaa.2004.07.001

Bicho N, Cascalheira J, Gonçalves C, Umbelino C, García Rivero D, André L (2017) Resilience, replacement and acculturation in the mesolithic/neolithic transition: the case of Muge, central Portugal. Quat Int 446:31–42. https://doi.org/10.1016/j.quaint.2016.09.049

Binford L (2001) Constructing frames of reference: an analytical method for archaeological theory building using ethnographic and environmental data sets. University of California Press, Berkeley

Bird DW, O'Connell JF (2006) Behavioral ecology and archaeology. J Archaeol Res 14:143–188. https://doi.org/10.1007/s10814-006-9003-6

Bonner N (1981) Southern fur seals. Arctocephalus (Geoffroy Saint-Hilarie and Cuvier, 1826). In: Ridgway SH, Harrison RJ (eds) Handbook of marine mammals, vol 1: the walrus, sea lions, fur seals and sea otter. Academic Press, London, pp 161–208

Boomert A, Bright AJ (2007) Island archaeology: in search of a New Horizon. Island Stud J 2(1):3–26. https://doi.org/10.24043/isj.197

Borrero LA (2011) Los cazadores orientales de Tierra del Fuego. In: Zangrando AF, Vázquez MM, Tessone A (eds) Los cazadores-recolectores del extremo oriental fueguino. Arqueología de Península Mitre e Isla de los Estados. Sociedad Argentina de Antropología, Buenos Aires, pp 287–298

Borrero LA, Martin FM (2023) Fragmented Records. Fuego-Patagonian Hunter-Gatherers and Archaeological Change. In: Wardle JD, Hitchcock RK, Schmader M, Yu P-L (eds) Archaeology on the threshold. Studies in the Processes of Change. University Press of Florida, Gainesville, pp 68–88

Borrero LA, Morello F, San Román M (2020) Circulación de bienes, uso del espacio interior y espacios programados en los archipiélagos de Fuego-Patagonia en tiempos recientes. Magallania 48(2):71–98. https://doi.org/10.4067/S0718-22442020000200071

Bove G (1883) Expedición Austral Argentina. Informes preliminares presentados a SSEE los Ministerios del Interior y de Guerra y Marina de la República Argentina. Instituto Geográfico Argentino, Buenos Aires

Boyd R, Richerson PJ, Henrich J (2011) The cultural niche: why social learning is essential for human adaptation. PNAS 108(supp 2):10918–10925. https://doi.org/10.1073/pnas.1100290108
Bradtmöller M, Grimm S, Riel-Salvatoreet J (2017) Resilience theory in archaeological practice—an annotated review. Quat Int 446:3–16. https://doi.org/10.1016/j.quaint.2016.10.002
Bridges T (1866) Letter published in a voice for South America XIII, London
Bridges T (1869) South American missionary magazine III, London
Bridges T (1874) South American missionary magazine VIII, London
Bridges T (1876) South American missionary magazine X, London
Bridges T (1880) South American missionary magazine XIV, London
Bridges T (1881) South American missionary magazine XV, London
Bridges T (1884) South American missionary magazine XVIII, London
Bridges T (1886) South American missionary magazine XX, London
Broodbank C (2008) Not waving but drowning. J Island Coast Archaeol 3:72–76. https://doi.org/10.1080/15564890801990771
Broughton J (1994) Late Holocene resource intensification in the Sacramento Valley, California: the vertebrate evidence. J Archaeol Sci 21:501–514. https://doi.org/10.1006/jasc.1994.1050
Broughton JM, Cannon MD, Bartelink EJ (2010) Evolutionary ecology, resource depression, and niche construction theory: applications to Central California hunter-gatherers and Mimbres-Mogollon agriculturalists. J Archaeol Method Theory 17:371–421. https://doi.org/10.1007/s10816-010-9095-7
Butto A, Fiore D (2020) Ritmos de cambio en la vestimenta y ornamentos de la sociedad Yagán de Tierra del Fuego (siglos XIX y XX). Anu Esc Hist Virtual 18:158–187
Butzer K (1982) Archaeology as human ecology. Cambridge University Press, Cambridge
Byrd J (1997) The analysis of diversity in archaeological faunal assemblages: complexity and subsistence strategies in the Southeast during the middle Woodland period. J Anthropol Archaeol 16:49–72. https://doi.org/10.1006/jaar.1997.0304
Cannon MD (1999) A mathematical model of the effects of screen size on zooarchaeological relative abundance indices. J Archaeol Sci 26:205–214. https://doi.org/10.1006/jasc.1998.0317
Cannon MD (2001) Archaeofaunal relative abundance, sample size, and statistical methods. J Archaeol Sci 28:185–195. https://doi.org/10.1006/jasc.2000.0558
Chapman A (1986) Los Selk´nam. La vida de los onas. Emecé, Buenos Aires
Chapman A (1987) La Isla de los Estados en la prehistoria. Primeros datos arqueológicos. Eudeba, Buenos Aires
Chapman A, Hester TR (1973) New data on the archaeology of the Haush: Tierra del Fuego. J Soc Américanistes LXII:185–208. https://www.jstor.org/stable/24605455
Coiazzi A (1914) Los indios del archipiélago fueguino. Rev Chil Hist Geogr 13:1–113
Crespo CM, Cardozo DG, Tessone A, Vázquez MM, Kisielinski C, Arencibia V, Tackney J, Zangrando AF, Dejean CB (2020) Distribution of maternal lineages in hunter-gatherer societies of the southern coast of Tierra del Fuego. Argentina. Am J Phys Anthropol 173(4):709–720. https://doi.org/10.1002/ajpa.24107
De Brosses C (1756) Historie des Navigations aux Terres Australes. Durand, Paris
Erlandson JM (2001) The Archaeology of aquatic adaptations: paradigms for a New Millennium. J Archaeol Res 9:287–350. https://doi.org/10.1023/A:1013062712695
Erlandson J (2008) Isolation, interaction, and island archaeology. J Island Coast Archaeol 3:83–86. https://doi.org/10.1080/15564890801999939
Erlandson JM, Rick TC, Braje TJ, Steinberg A, Vellanoweth RL (2008) Human impacts on ancient shellfish: a 10,000 year record from San Miguel Island. California. J Archaeol Sci 35(8):2144–2152. https://doi.org/10.1016/j.jas.2008.01.014
Estévez J, Vila A (1995) Encuentros en los conchales fueguinos. Treballs D'Etnoarqueologia 1. CSIC, Universidad Autónoma de Barcelona, Barcelona
Estévez J, Vila A (2006) Variability in the lithic and faunal record through 10 reoccupations of a XIX century Yamana Hut. J Anthropol Archaeol 25(4):408–423. https://doi.org/10.1016/j.jaa.2006.03.007

Estévez J, Vila A (2007) Twenty years of ethnoarchaeological research in Tierra del Fuego: some thoughts for European shell-midden archaeology. In: Craig OE, Bailey GN (eds) Milner N. Oxbow Books, Oxford, pp 183–195

Esteves JL, Ciocco NF, Colombo JC, Freije H, Harris G, Iribarne O, Isla I, Pascual MS, Penchaszadeh PE, Rivas AL, Santinelli N (2000) The Argentine sea: the southeast South American shelf marine ecosystem. In: Sheppard CRC (ed) Seas at the Millenium: an environmental evaluation. Pergamon, New York, pp 749–771

Falabella V, Campagna C, Croxall J (2009) Atlas del Mar Patagónico. Especies y espacios. Wildlife Conservation Society & BirdLife International, Buenos Aires

Fitz Roy R (2016 [1839]) Los viajes del Beagle. Informes de la Segunda Expedición (1831–1836). Colección Reservada del Museo del Fin del Mundo. Eudeba, Buenos Aires

Fitzhugh W (1997) Biogeographical archaeology in the Eastern North American Arctic. Hum Ecol 25(3):385–418. https://doi.org/10.1023/A:1021819509181

Fitzhugh B (2001) Risk and invention in human technological evolution. J Anthropol Archaeol 20:125–167. https://doi.org/10.1006/jaar.2001.0380

Fitzhugh B, Gjesfjeld EW, Brown WA, Hudson MJ, Shaw JD (2016) Resilience and the population history of the Kuril Islands, Northwest Pacific: a study in complex human ecodynamics. Quat Int 419:165–193. https://doi.org/10.1016/j.quaint.2016.02.003

Fitzhugh B, Brown WA, Misarti N (2020) Archaeological demography of the North Pacific Rim: modeling crashes, climates, and social dynamics. In: Krupnik I, Crowell AL (eds) Arctic crashes: people and animals in the changing north. Smithsonian Scholarly Press, Washington DC, pp 43–60

Fitzpatrick SM, Anderson A (2008) Islands of isolation: archaeology and the power of aquatic perimeters. J Isl Coast Archaeol 3(1):4–16. https://doi.org/10.1080/15564890801983941

Furlong CW (1917) Tribal distribution and settlements of the Fuegian, comprising nomenclature, etymology, philology, and populations. Geogr Rev III 3:169–187

Gardiner AW (1857) The voice of pity for South America IV:185-188. Wertheim, Macintosh and Hunt, London.

Gardiner AW (1858) The voice of pity for South America V:219-228. Wertheim, Macintosh and Hunt, London.

Grattan J, Torrence R (2007) Beyond gloom and doom: the long-term consequences of volcanic disasters. In: Grattan J, Torrence R (eds) Living under the shadow: the cultural impacts of volcanic eruptions. Routledge, New York, pp 1–18

Grayson DK (1984) Quantitative zooarchaeology: topics in the analysis of archaeological faunas. Academic Press, Orlando

Gusinde M (1986 [1937]) Los indios de Tierra del Fuego. Los Yámana (Tomo II). Centro Argentino de Etnología Americana (CONICET), Buenos Aires

Hawkesworth J (1773) An account of the voyages… performed by Commodore Byron, Captain Carteret, Captain Wallis, and Captain Cook… drawn up from the journals which were kept by the several commanders, and from the papers of Joseph Banks. W Strahan & T Cadell, London

Hill MO (1973) Diversity and evenness: a unifying notation and its consequences. Ecology 54:427–432. https://doi.org/10.2307/1934352

Hill KR, Wood BM, Baggio J, Hurtado AM, Boyd RT (2014) Hunter-Gatherer inter-band interaction rates: implications for cumulative culture. PLoS ONE 9(7):e102806. https://doi.org/10.1371/journal.pone.0102806

Horwitz V (1990) Maritime settlement patterns in Southeastern Tierra del Fuego (Argentina). PhD Dissertation, University of Kentucky

Horwitz V (1993) Maritime settlement patterns: the case from Isla de los Estados (Staten Island). Arqueol Contemp 4:163–176

Horwitz V, Weissel M (2011) Arqueología de Isla de los Estados: La frontera de la abundancia. In: Zangrando AF, Vázquez M, Tessone A (eds) Los cazadores recolectores del extremo oriental fueguino. Arqueología de Península Mitre e Isla de los Estados. Sociedad Argentina de Antropología, Buenos Aires, pp 143–170

Hudson MJ, Aoyama M, Hoover KC (2009) Navigating Hunter-Gatherer resilience: networks and insularity in the prehistory of the Ryukyu Islands. Mémoires De La Société Finno-Ougrienne 265:49–66

Hyades P (1884) Notes hygiéniques et médicales sur les Fuégiens de l'Archipel du Cap Horn. Revue d´Hygiéne et de Police Sanitaire 6(7):550–590

Hyades P, Deniker J (2007 [1891]) Los Indígenas según Hyades y Deniker. In: Legoupil D, Prieto A (eds) Etnografía de los indios Yaghan en la Misión Científica. Ediciones Universidad de Magallanes—Instituto Francés de Estudios Andinos, Punta Arenas, pp 69–329

Janssen MA, Bodin Ö, Anderies JM, Elmqvist T, Ernstson H, McAllister RRJ, Olsson P, Ryan P (2006) A network perspective on the resilience of social-ecological systems. Ecol Soc 11(1):15. http://www.ecologyandsociety.org/vol11/iss1/art15/

Jones TL, Richman JR (1995) On mussels: *Mytilus californianus* as a prehistoric resource. North Am Archaeol 16(1):33–58. https://doi.org/10.2190/G5TT-YFHP-JE6A-P

Klein RG, Steele TE (2013) Archaeological shellfish size and later human evolution in Africa. PNAS 110(27):10910–10915. https://doi.org/10.1073/pnas.1304750110

Kochi S (2023) El nicho ecológico de los cazadores-recolectores del canal Beagle, Península Mitre e Isla de los Estados durante el Holoceno tardío y sus implicancias socioculturales. PhD Dissertation, Universidad de Buenos Aires

Laland KN, Odling-Smee J, Feldman MW (2001) Cultural niche construction and human evolution. J Evol Biol 14:22–33. https://doi.org/10.1046/j.1420-9101.2001.00262.x

Lefèvre, C (1993–94) Las aves en los yacimientos del Cabo de Hornos y del Seno Grandi. Anal Inst Patag Ser Cs Hum 22:123–136

Legoupil, D (1993–94) El archipiélago del Cabo de Hornos y la costa sur de la isla Navarino: poblamiento y modelos económicos. Anal Inst Patag Ser Cs Hum 22:101–121

Legoupil D (1995) Des indigènes au Cap Horn: conquête d'un territoire et modèle de peuplement aux confins du continentsud-américain. J Soc Amér 81:9–45. https://doi.org/10.3406/jsa.1995.1582

Legoupil D, Fontugne M (1997) El poblamiento marítimo de los archipiélagos de Patagonia: núcleos antiguos y dispersión reciente. Anal Inst Patag Ser Cs Hum 25:75–87

Lista R (1887) Viaje al País de los Onas, Tierra del Fuego. Establecimiento Tipográfico de Alberto Nuñez, Buenos Aires

Lloris D, Rucabado J (1991) Ictiofauna del canal Beagle (Tierra del Fuego): aspectos ecológicos y análisis biogeográfico. Instituto Español de Oceanografía (Publicación Especial 8), Madrid

Lothrop SK (1928) The indians of Tierra del Fuego. Museum of the American Indian, Heye Foundation, New York

Lyman L (2003) Pinniped behaviour, foraging theory, and the depression of metapopulation and nondepression of a local population on the southern Northwest Coast of North America. J Anthropol Archaeol 22:376–388. https://doi.org/10.1016/S0278-4165(03)00022-9

Mackie Q (2001) Settlement archaeology in a Fjordland archipelago: Network analysis, social practice and the built environment of Western Vancouver Island, British Columbia, Canada since 2000 BP. BAR International Series 926, Oxford

Mannino MA, Thomas KD (2002) Depletion of a resource? The impact of prehistoric human foraging on intertidal mollusc communities and its significance for human settlement, mobility and dispersal. World Archaeol 33(3):452–474. https://doi.org/10.1080/00438240120107477

Martial LF (2007 [1888]) Los indígenas según Martial. In: Legoupil D, Prieto, A (eds) Etnografía de los indios Yaghan en la Misión Científica. Ediciones Universidad de Magallanes—Instituto Francés de Estudios Andinos, Punta Arenas, pp 17–68

Martinoli MP (2018) Modalidades de explotación, procesamiento y consumo de pinnípedos en la margen meridional de Tierra del Fuego. PhD Dissertation, Universidad de Buenos Aires

Martinoli MP, Zangrando AF (2024) Estrategias de aprovisionamiento de otáridos en la margen meridional de Tierra del Fuego durante el Holoceno. Lat Am Antiq 35(4):835–852. https://doi.org/10.1017/laq.2023.60

Maximiano Castillejo A (2016) From virtual survey to real prospection: Kaw'esqar mobility in the Fuego-Patagonia seascape across terrestrial passages. Quat Int 435:114–127. https://doi.org/10.1016/j.quaint.2015.10.078

Milano VN, Grandi M, Schiavini AM, Crespo E (2020) Recovery of South American fur seals from Fuegian Archipelago (Argentina). Mar Mammal Sci 36:1022–1032. https://doi.org/10.1111/mms.12686

Nelson MC, Hegmon M, Kintigh KW, Kinzig AP, Nelson BA, Anderies JM, Abbott DA, Spielmann KA, Ingram SE, Peeples MA, Kulow S, Strawhacker CA, Meegan C (2012) Long-term vulnerability and resilience: three examples from archaeological study in the Southwestern United States and Northern Mexico. In: Cooper J, Sheets P (eds) Surviving sudden environmental change. Understanding Hazards, Mitigating Impacts, Avoiding Disasters. University Press of Colorado, Colorado, pp 197–221

Nye JW (2019) Prehistoric and Modern Ecological Dynamics in Southern South American Marine Food Webs. PhD Dissertation, University of California

Nye JW, Zangrando AF, Martinoli MP, Fogel ML (2020) Temporal and population trends in human exploited pinnipeds from Tierra del Fuego. Palaeogeogr Palaeoclimatol Palaeoecol 55:4109804. https://doi.org/10.1016/j.palaeo.2020.109804

Orquera LA (2002) The late nineteenth-century crisis in the survival of the Magellan-Fueguian Litoral Natives. In: Briones C, Lanata JL (eds) Archaeological and anthropological perspectives on the native peoples of Pampa, Patagonia, and Tierra del Fuego to the Nineteenth Century. Bergin & Garvey, London, pp 145–158

Orquera LA, Piana EL (1986–87) Composición tipológica y datos tecnomorfológicos y tecnofuncionales de los distintos conjuntos arqueológicos del sitio Túnel I (Tierra del Fuego). Rel Soc Arg Antropol XVII:201–239

Orquera LA, Piana EL (1996) El sitio Shamakush I (Tierra del Fuego, República Argentina). Rel Soc Arg Antropol XXI:215–265

Orquera LA, Piana EL (1999a) Arqueología de la región del canal Beagle (Tierra del Fuego, República Argentina). Sociedad Argentina de Antropología, Buenos Aires

Orquera LA, Piana EL (1999b) La vida material y social de los Yámana. Eudeba, Buenos Aires

Orquera LA, Piana EL (2001) Composición de Conchales de la costa del canal Beagle (Tierra del Fuego, República Argentina) (Segunda parte). Rel Soc Arg Antropol XXVI:345–368

Orquera LA, Sala A, Piana EL, Tapia A (1977) Lancha Packewaia. Arqueología de los Canales Fueguinos. Temas de Arqueología. Huemul, Buenos Aires

Parkington J, Fisher JW, Kyriacou K (2013) Limpet gathering strategies in the later stone age along the Cape west coast. South Africa. J Isl Coast Archaeol 8(1):91–107. https://doi.org/10.1080/15564894.2012.756084

Payró, RJ (1898) La Australia argentina (excursión periodística a las costas patagónicas, Tierra del Fuego e Isla de los Estados). La Nación, Buenos Aires

Pérez SA (2020) La explotación de moluscos intermareales durante el Holoceno Tardío en la localidad de Heshkaia, bahía Moat, sur de Tierra del Fuego. Intersecc Antropol 21(1):43–56. https://doi.org/10.37176/iea.21.1.2020.529

Piana EL, Orquera LA (2010) Shell midden formation at the Beagle Channel (Tierra del Fuego, Argentine). In: Calado D, Baldia M, Boulanger M (eds) Monumental questions: prehistoric megaliths, mounds and enclosures British archaeological reports international series 2122. Archaeopress, Oxford, pp 263–273

Piana EL, Vázquez MM, Rúa N (2004) Mischiuén I. Primeros resultados de una excavación de rescate en la costa norte del canal Beagle. In: Civalero MT, Fernández PM, Guráieb AG (eds) Contra viento y marea. Arqueología de Patagonia. INAPL, Buenos Aires, pp 815–832

Prieto A, Chevallay D, Ovando D (2000) Los pasos de indios en Patagonia Austral. In: Carballo MF, Espinosa S, Belardi JB, De Nigris M, Fernández P, Gómez Otero J, Goñi R (eds) Desde el País de los Gigantes. Universidad Nacional de la Patagonia Austral, Río Gallegos, Perspectivas arqueológicas en Patagonia, pp 87–94

Roberts P, Stewart BA (2018) Defining the 'generalist specialist' niche for Pleistocene *Homo sapiens*. Nat Hum Behav 2:542–550. https://doi.org/10.1038/s41562-018-0394-4
Ross JC (1847) A voyage of discovery and research in the southern and antarctic regions during the years 1839–43. John Murray, London.
Scheinsohn V (1997) Explotación de materias primas óseas en la Isla Grande de Tierra del Fuego. PhD Dissertation, Universidad de Buenos Aires
Schiavini AM (1993) Los lobos marinos como recurso para cazadores-recolectores marinos: el caso de Tierra del Fuego. Lat Am Antiq 4:346–366. https://doi.org/10.2307/972072
Segers P (1891) Hábitos y costumbres de los indios Aonas. Bol Inst Geogr Argent 12:56–82
Segura IJA (1973) Don Manuel Lefrant en la bahía fueguina del Buen Suceso. Karukinka 4:3-8
Spegazzini C (1882) Costumbres de los habitantes de la Tierra del Fuego. An Soc Cienc Argent XIV:159–181
Snow WP (1857) A two year's cruise off Tierra del Fuego, the Falkland Islands, Patagonia and in the River Plate; a narrative of life in the southern seas. Longman, London.
Snow WP (1861) A few remarks on the wild tribes of Tierra del Fuego from personal observation. Trans Ethn Soc London I:261–267
Snow WP (1864) A cruise among the Fuegians. Harpers Mag XXVIII (164):160–167
Stiner MC, Kuhn SL (2006) Changes in the 'connectedness' and resilience of paleolithic societies in mediterranean ecosystems. Hum Ecol 34:693–712. https://doi.org/10.1007/s10745-006-9041-1
Tafuri MA, Zangrando AF, Tessone A, Kochi S, Moggi Cecchi J, Di Vincenzo F, Profico A, Manzi G (2017) Dietary resilience among hunter-gatherers of Tierra del Fuego: Isotopic evidence in a diachronic perspective. PLoS ONE 12(4):e0175594. https://doi.org/10.1371/journal.pone.0175594
Tivoli AM, Zangrando AF (2011) Subsistence variations and landscape use among maritime hunter-gatherers. A zooarchaeological analysis from the Beagle Channel (Tierra del Fuego, Argentina). J Archaeol Sci 38:1148–1156. https://doi.org/10.1016/j.jas.2010.12.018
Vázquez MM, Zangrando AF, Tessone A, Ceraso A (2011) Investigaciones arqueológicas recientes en la costa suroriental de Península Mitre. In: Zangrando AF, Vázquez MM, Tessone A (eds) Los cazadores-recolectores del extremo oriental fueguino. Arqueología de Península Mitre e Isla de los Estados. Sociedad Argentina de Antropología, Buenos Aires, pp 203–203
Vidal H (2011 [1985]) El conocimiento de las culturas aborígenes del sudeste de la Isla Grande hasta 1983. In: Zangrando AF, Vázquez MM, Tessone A (eds) Los cazadores-recolectores del extremo oriental fueguino. Arqueología de Península Mitre e Isla de los Estados. Sociedad Argentina de Antropología, Buenos Aires, pp 65–113
Vidal H, Winograd A (1986) El factor humano. Modificaciones antropogénicas en la paleoecología de otáridos en Tierra del Fuego. In: Anais da 2ª Reuniâo de trabalho de especialistas em mamíferos aquáticos da America do Sul. Fundação Brasileira para a Conservação da Naturaleza, Río de Janeiro, pp 96–99
Weddell J (2006 [1825]) Un viaje hacia el Polo Sur realizado en los años 1822–1824. Colección Reservada del Museo del Fin del Mundo, Eudeba, Buenos Aires.
Walker BH, Gunderson LH, Kinzig AP, Folke C, Carpenter SR, Schultz L (2006) A handful of heuristics and some propositions for understanding resilience in social-ecological systems. Ecol Soc 11(1):13. http://www.ecologyandsociety.org/vol11/iss1/art13/
Winterhalder B, Smith EA (1992) Evolutionary ecology and the social sciences. In: Smith EA, Winterhalder B (eds) Evolutionary ecology and human behavior. Aldine de Gruyter, Nueva York, pp 3–23
Yesner D (1990) Fuegians and other Hunter-Gatherers of the subantarctic region: 'cultural devolution' reconsidered. In: Meehan B, White N (eds) Hunter-Gatherer demography. Past and Present, Oceania Monograph 39. University of Sydney, Sydney, pp 1–22
Zangrando AF (2009a) Historia evolutiva y subsistencia de cazadores-recolectores marítimos de Tierra del Fuego. Sociedad Argentina de Antropología, Buenos Aires

Zangrando AF (2009b) Is fishing intensification a direct route to hunter-gatherer complexity? A case study from the Beagle Channel region (Tierra del Fuego, southern South America). World Archaeol 41(4):589–608. https://doi.org/10.1080/00438240903363848

Zangrando AF (2010) Coastal archaeology and hunter-gatherers in the south-eastern of Tierra del Fuego. J Isl Coast Archaeol 5(2):288–291. https://doi.org/10.1080/15564894.2010.487360

Zangrando AF (2018) Poblamiento temprano y arqueología de costas en Patagonia y Tierra del Fuego: vacío de información, preconceptos y perspectivas. Intersecc Antropol 19:63–97

Zangrando AF, Tessone A, Vázquez M (2009) El uso de espacios marginales en el Archipiélago Fueguino: implicaciones de la evidencia arqueológica de Bahía Valentín. In: Salemme M, Santiago F, Álvarez MR, Piana EL, Vázquez MM, Mansur E (eds) Arqueología de Patagonia: una mirada desde el último confín. Utopías, Ushuaia, pp 47–62

Zangrando AF, Borrazo K, Tivoli A, Alunni D, Martinoli MP (2014a) El sitio Heshkaia 35: nuevos datos sobre la arqueología de Moat (Tierra del Fuego, Argentina). Rev Mus Antropol 7:11–24. https://doi.org/10.31048/1852.4826.v7.n1.9090

Zangrando AF, Panarello H, Piana EL (2014b) Zooarchaeological and stable isotopic assessments on pinniped-human relations in the Beagle Channel (Tierra del Fuego, Southern South America). Int J Osteoarchaeol 24(2):231–244. https://doi.org/10.1002/oa.2352

Zangrando AF, Ponce JF, Martinoli MP, Montes A, Piana E, Vanella F (2016) Paleogeographic changes drove prehistoric fishing practices in the Cambaceres Bay (Tierra del Fuego, Argentina) during the middle and late Holocene. Environ Archaeol:1–11. https://doi.org/10.1080/14614103.2015.1130888

Zangrando AF, Tivoli AM, Alunni DV, Pérez SA, Martinoli MP, Pinto Vargas G (2021) Exploring shell midden formation through tapho-chronometric tools: a case study from Beagle Channel, Argentina. Quat Int 584:33–43. https://doi.org/10.1016/j.quaint.2020.04.050

Zangrando AF, Kochi S, Nye JW, Tessone A, Riccialdelli L, Tivoli AM, Martinoli MP, Pérez SA, Fogel ML (2022a) Long-term interaction between Hunter-Gatherer-Fisher societies and marine resources in the southern tip of South America applications of stable isotopes in zooarchaeology. In: Sharpe A, Krigbaum J (eds) Isotope research in zooarchaeology: methods, applications, and advances. University Press of Florida, Florida, pp 15–43

Zangrando AF, Tivoli AM, Ponce JF, Alunni DA, Fernández Ropero MC, Martinoli MP (2022b) Colonización humana de Tierra del Fuego: contextos arqueológicos pre-transgresivos en la costa norte del Canal Beagle. Chungará 54(3):557–577. https://doi.org/10.4067/S0717-73562022005000401

Zangrando, AF, Martinoli MP, Tessone A, Vázquez MM, Alunni DV, Franch Bach A, Hernández Herrero O, Fernández Ropero MC, Pinto Vargas G, Tivoli AM (2024) El uso humano de espacios externos en el archipiélago de Tierra del Fuego durante el Holoceno: el caso de Isla de los Estados. Bol Soc Chil Arqueol 57:85–119. https://doi.org/10.56575/BSCHA.05700240843

Chapter 2
Exploring Cultural Niche Dynamics in Time: Coastal Settlements, Chronology and Technology in the Southern Coast of Tierra del Fuego

Atilio Francisco J. Zangrando, Juan F. Ponce, Dánae Fiore, Oriana Hernandez Herrero, and Anna Franch Bach

Abstract This chapter explores the cultural niche dynamics of hunter-gatherer-fisher populations along the southern coast of Tierra del Fuego during the Holocene. Integrating palaeoenvironmental data, demographic reconstructions, archaeological site distributions, technological assemblages and genetic evidence, this study examines how human groups adapted to environmental variability and shifting social interactions over time. Radiocarbon-based demographic models reveal periods of significant population growth, especially between 7500 and 6300 and around 500 cal yr BP, interrupted by phases of decline, notably between 4800 and 2000 cal yr BP. Technological transitions, such as the adoption of bifacial projectile points and shifts in bone tool diversity, align with demographic and ecological changes. Spatial analysis

---

A. F. J. Zangrando (✉) · J. F. Ponce · O. Hernandez Herrero · A. Franch Bach
CADIC-CONICET, Ushuaia, TDFeIAS, Argentina
e-mail: fzangrando@conicet.gov.ar

J. F. Ponce
e-mail: jfponce@conicet.gov.ar

O. Hernandez Herrero
e-mail: orianaa.hernandez97@gmail.com

A. Franch Bach
e-mail: Anna.Franch@uab.cat

A. F. J. Zangrando · D. Fiore
FFyL-UBA, CABA, Argentina
e-mail: danae_fiore@yahoo.es

J. F. Ponce
ICPA-UNTDF, Ushuaia, TDFeIAS, Argentina

D. Fiore
AIA-CONICET, CABA, Argentina

A. Franch Bach
UAB, Barcelona, Spain

A. F. J. Zangrando et al. (eds.), *Archaeology of the South Coast of Tierra del Fuego*, The Latin American Studies Book Series, https://doi.org/10.1007/978-3-032-09487-2_2

of lithic raw material circulation—especially green obsidian—demonstrates long-distance interaction networks and fluctuating connectivity across the archipelago. The study further examines genetic data to understand inter-group relations and cultural transmission mechanisms. This work highlights how cultural niche construction in this sub-Antarctic environment was shaped by resilience strategies, technological innovations and the evolution of social networks in response to dynamic climatic and demographic pressures.

2.1 Introduction

As discussed in the previous chapter, different natural and social forces influence cultural niche construction in hunter-gatherer and fisher (HGF) populations. While niche construction is a universal process exhibited by all living organisms (Odling-Smee et al. 2003), humans modify their social and natural environments predominantly through cultural processes (Laland and O'Brien 2011). In this context, the transmission of knowledge via social interactions serves as a crucial mechanism in shaping the human niche.

This chapter outlines a framework for understanding population and social dynamics, exploring variations in the archaeological record of HGF populations that inhabited the southern coast of Tierra del Fuego. These variations are interpreted as strategic responses to differing levels of environmental variability and interaction costs. To achieve this, we first establish the geographical framework and highlight key palaeoclimatic events of the Holocene. We then examine various population parameters, including palaeodemographic trends, the formation and spatial distribution of archaeological sites and interaction dynamics, as evidenced by social networks, technological innovations and regional gene flow patterns.

2.2 Geographic Setting

The landscape of the Fuegian Archipelago can be divided into two main regions: the Fuegian Andes to the south and west and the high plains to the north (Fig. 2.1). The Fuegian Andes constitute a mountain range with NW–SE orientation along the Pacific margin and W–E orientation in the southern part of the archipelago, exhibiting a decrease in elevation from south to north and from west to east. These are the lowest Andean peaks in Patagonia. The highest elevations are found at Mount Darwin (2488 m a.s.l.) in the Darwin Range. Isla de los Estados represents the southeastern end of the Fuegian Andes. The landscape of Isla Grande de Tierra del Fuego was modified by several outlet glaciers that descended from the Darwin Range.

The climate of the Fuegian Archipelago is primarily influenced by its location at mid-high latitudes and the dominant Southern Westerly Winds (SWW) belt, in the path of eastward-moving cyclones not far from Antarctic ice. The latitudinal

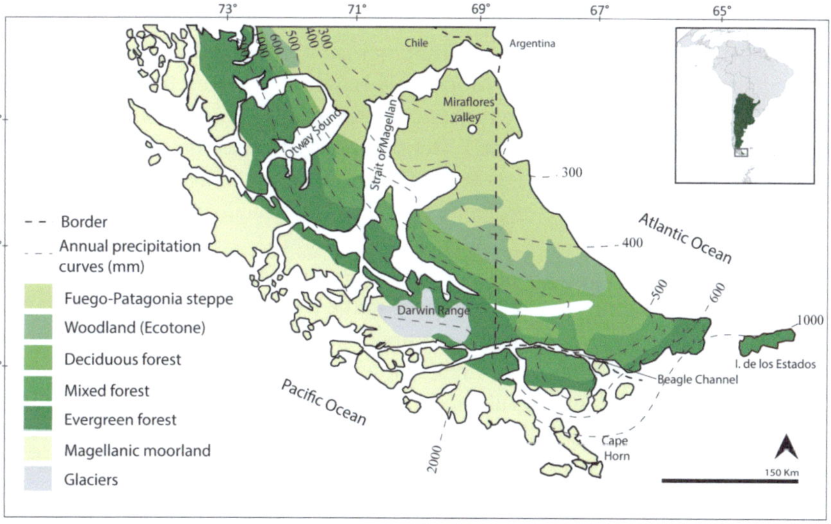

Fig. 2.1 Map of phytogeographic units and annual precipitation curves (in mm), modified from Borromei et al. (2014)

position and strength of the SWW are controlled by the intensity and latitudinal changes of the subtropical high-pressure cells in the Pacific and the circum-Antarctic low-pressure corridor (Tuhkanen 1992). The regional climate, with an annual mean temperature of 5 °C, is highly oceanic in the western and southern sectors of the archipelago and increasingly continental toward the east and north. The most frequent winds are from the south and southwest (Rabassa et al. 2000; Coronato et al. 2008). Annual precipitation shows a strong gradient, decreasing from around 2000 mm in the southwest to less than 300 mm in the northeast (Tuhkanen 1992). A west-to-east annual precipitation gradient increase is also evident with 1000 mm at the easternmost tip of Tierra del Fuego and close to 1500 mm at Isla de los Estados (Ponce et al. 2011).

The vegetation composition reflects the pronounced trans-Andean climatic and topographic gradients (Moore 1983; Tuhkanen 1992). The sub-Antarctic deciduous forest is represented by two southern beech species, *Nothofagus pumilio* (lenga) and *Nothofagus antarctica* (Antarctic beech or ñire), which grow from sea level to an average altitudinal limit of 550–600 m above sea level, and become dominant where annual precipitation exceeds 450 mm (Pisano 1977; Moore 1983). To the south and west of Tierra del Fuego, annual rainfall exceeds 700 mm and the Subantarctic Evergreen Forest develops, dominated by *Nothofagus betuloides* (Magellan's beech or guindo), accompanied by *Drimys winteri* (Winter's bark or canelo), *Maytenus magellanica* (Magellan's mayten or maitén) and abundant ferns and mosses (Moore 1983). Initially, *N. betuloides* intermingles with *N. pumilio,* forming a mixed evergreen-deciduous forest that represents an ecotonal transition between both communities. (Moore 1983). The Magellanic moorland is found beyond the forest along the exposed outer coast under conditions of higher rainfall (>1500 mm), strong winds and poor drainage. It consists of a mosaic of barren rock, marginal grassland, peatland, scrub and patches of evergreen forest (Moore 1983). Above the tree line in the Fuegian Cordillera, the Andean tundra consists of cushion plants, dwarf shrub heaths and grassland communities (Pisano 1977; Heusser 2003). Reflecting its exposure to high rainfall and oceanic influence, Isla de los Estados is characterized by lush vegetation dominated by Subantarctic Evergreen Forest and Magellanic moorland (Ponce et al. 2017).

Along the southern coast of Tierra del Fuego, three distinctive ecological zones can be distinguished for human subsistence in the past (Tivoli and Zangrando 2011, Chap. 3). In the terrestrial ecosystem, an important part of biotic energy is absorbed by the formation of woody structures. The supply of carbohydrates and proteins of plant origin is very small (Tuhkanen 1992), so there is little animal life in the forests. Few mammal and bird species inhabit the terrestrial domain, among which guanacos (*Lama guanicoe*) were the most important resource for HGF populations. Aquatic environments, such as rivers and lakes, lack native animal species important for human subsistence. The coastal ecozone divides marine and terrestrial systems. The coastal margins provide the greatest amount of biomass resources for human subsistence and present the most favorable conditions in terms of accessibility and predictability for resource procurement. Two species of otariids occur in this ecozone: sea lions (*Otaria flavescens*) and fur seals (*Arctocephalus australis*). There is also an important biodiversity of birds in the coastal areas, including different species of penguins, cormorants and anatidae. Several of these species are gregarious. Numerous species of fish are represented in the shallow waters. Most of them belong to the family *Nototheniidae*, in which small species predominate. An important diversity of mollusks is also present in the intertidal. As far as purely pelagic mammals are concerned, there are several species of cetaceans: a variety of dolphins, baleen whales, porpoises and sperm whales. Among birds, there are different species of petrels, shearwaters and albatrosses. All these birds usually inhabit the offshore and outer islands of the Fuegian Archipelago, although some of these species enter into the Beagle Channel. In deep waters, the fish community includes key resources for the marine ecosystem, such as the Fuegian sardine (*Sprattus fueguensis*). Several species of predatory fish are also present, such as hake (*Macruronus magellanicus*, *Merluccius hubbsi* and *Merluccius australis*) and snoek (*Thyrsites atun*).

2.3 Main Paleoenvironment Events

In a previous study, we identified five palaeoenvironmental events that significantly impacted the biomes and human populations of Tierra del Fuego during the Holocene (Fernández et al. 2020).

2.3.1 Forest Expansion

The Late Glacial-Holocene transition was characterized by the *Nothofagus* expansion, from multiple ice-free refuges (Premoli et al. 2010) located within the current extension of the forest, while the herbaceous-shrub communities remained. Patches of *Nothofagus* forest and steppe vegetation dominated the southern Tierra del Fuego landscape during this time period (Coronato et al. 2022). McCulloch et al. (2020) reported a sustained period of drier conditions between 9700 and 6000 cal yr BP on Navarino Island, which likely led to a contraction of forest communities during an otherwise general trend of forest expansion. According to these authors, the reduction in effective humidity suggests a weaker influence of the Southern Westerly Winds (SWW), due to the extreme migration of the wind belt toward polar latitudes. In La Correntina peat bog Musotto et al. (2016) registered a decreased frequency of *Nothofagus* between ca. 8600 and 6500 cal yr BP (Figs. 2.2 and 2.4).

The gradual expansion of forest communities in the south of Tierra del Fuego seem to coincide with the gradual decrease in temperature after the Holocene Thermal Maximum and with the low intensity of the SWW in these latitudes according to Björck et al. (2012). This would have provoked a gradual and oscillating migration of the SWW belt core from Antarctic latitudes toward the south of Tierra del Fuego (Coronato et al. 2022). In the south of the Argentinian sector of Isla Grande de Tierra del Fuego and Isla de los Estados, both the *Nothofagus* pollen and plant macrofossils records from peat bogs indicate an increase in effective humidity and cold conditions from ca. 6700 cal yr BP, which would have led to the expansion of the closed-canopy

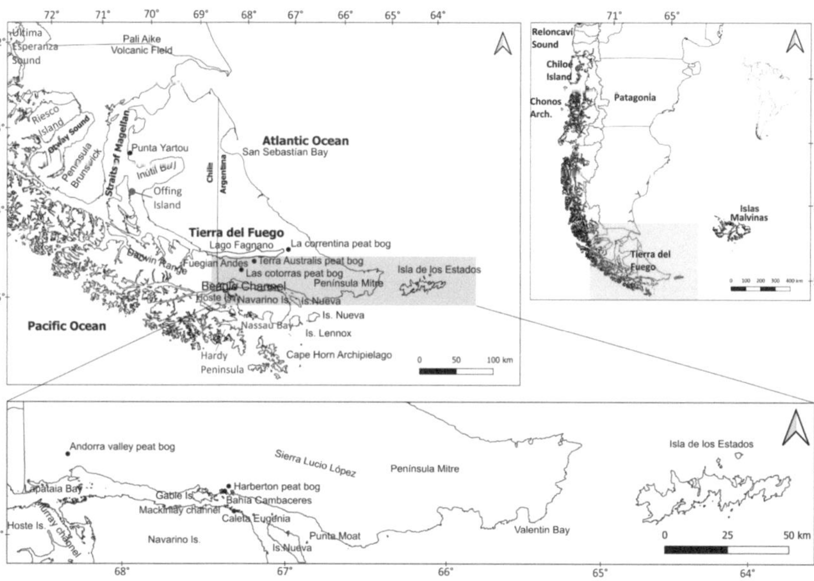

Fig. 2.2 Map showing geographical references and locations of peat bogs mentioned in the text

Nothofagus forest (Coronato et al. 2022). Pollen records from southern Tierra del Fuego show the highest development of the closed-canopy *Nothofagus* forest between ca. 5500 and 3000 cal yr BP under windy, humid and cold conditions. This evidence suggests that the SWW belt core could have been established at 54–55° S during this time interval (Ponce et al. 2011; Björck et al. 2012) with a maximum intensity ca. 3800 cal yr BP.

2.3.2 *Holocene Marine Transgression*

The geomorphological evolution of the Beagle Channel during the Holocene marine transgression remains a topic of debate. Based on the bathymetric thresholds identified for the Mackinlay and Murray channels, a fully marine environment near the Beagle Channel was proposed to have existed prior to 9100 cal yr BP (8200 ^{14}C yr BP) (Rabassa et al. 1986). Using morphological analysis, bathymetry and seismic profiles of the channel bed, Bujalesky (2011) suggested that the present-day Beagle Channel valley was rapidly inundated by the sea shortly after 11,000 cal yr BP. Mörner (1987, 1991) argued that differential tectonic uplift influenced the channel environment and that the region has not experienced significant glacio-isostatic deformation during the Holocene. A study carried out in the vicinity of Caleta Eugenia (Navarino Island) points to the onset of marine transgression 8700 cal yr BP (7925 ± 25^{14}C yr BP) in the eastern sector of the Beagle Channel (McCulloch et al. 2019). Another study along the northern coast of Isla Navarino presents two sea-level change curves for the area during the Holocene, one for the western sector and one for the central and eastern sector of the island (Björck et al. 2021). The curves indicate a rapid sea-level rise between 9000 and 7000–7500 cal yr BP, with peak highstands reaching approximately 7.5 m a. s. l. present sea level in the eastern sector between 7500 and 7000 cal yr BP, and up to 11 m in the western sector between 7000 and 6000 cal yr BP. These curves also show that a sea level similar to the present one was reached during the transgression between 8500 and 8700 cal yr BP. To the west of Ushuaia, in two sites located in Lapataia Bay (Lapataia Bay and Aserradero-Lapataia 2), marine sediments dated at 8478 cal yr BP and 8278 cal yr BP, respectively, were recorded (Borromei and Quattrocchio 2007; Candel et al. 2018). On the northern coast of the Beagle Channel, the altitudes of elevated sea levels associated with the maximum extent of this transgressive event, approximately 7300–6700 cal yr BP, range from 8–10 m a.s.l. in the western region to around 5.5 m a.s.l. in the eastern region (Rabassa et al. 1990; Gordillo et al. 1992, 1993; Bujalesky 1998, 2007; Zangrando et al. 2016). Toward the east, tectonic subsidence recorded on the southern coast of the Península Mitre (Rabassa et al. 2004) may have also affected Isla de los Estados, as indicated by the absence of elevated Holocene marine terraces on the island (Ponce et al. 2009).

Palaeogeographic models for the eastern end of the Beagle Channel, based on sea-level change curves proposed by Fleming et al. (1998) and Lambeck et al. (2014), suggest the existence of a paleo-bay east of Punta Moat. This bay was located near the final foothills of the Sierra Lucio López (on the southern coast of Tierra del

Fuego) and Isla Nueva, approximately 12,000 cal yr BP, when the sea level was about 50 m below its present level (Borromei et al. 2014). During this period, a paleolake is thought to have existed west of Punta Moat, which was subsequently inundated by seawater around 11,000 cal yr BP as the relative sea level rose to − 40 m. This inundation likely resulted in the formation of an extensive palaeo-fjord (Borromei et al. 2014).

2.3.3 Impact of the Hudson (H1) Volcanic Eruption

Several tephra layers, regionally distributed in extreme southern Patagonia, provide evidence of frequent explosive volcanic eruptions during Late Glacial and Holocene times (Stern 2008). Most notable in Tierra del Fuego is the identification of a green–brown tephra layer occurring in soil profiles, peatlands and alluvial and lacustrine deposits (Borromei et al. 2010, 2016; Menounos et al. 2013; Waldmann et al. 2014; Musotto et al. 2016, 2017). According to geochemical and petrological analyses, this tephra was derived from a large explosive eruption (H1) of Hudson Volcano (45° 54′S; 72° 58′W, 1905 m a.s.l.) and was regionally dated between 7960 and 7423 cal yr BP (Stern 2008). This volcano is located at the southern end of the Southern Andean Volcanic Zone (ZVS, 33°–46°S), > 900 km further northwest of the Beagle Channel. Middle Holocene tephra (H1) deposition in Tierra del Fuego (Unkel et al. 2010) probably could have caused a change in plant composition. In pollen records from the sites Cañadón del Toro (Borromei et al. 2016), Terra Australis (Musotto et al. 2017), La Correntina (Musotto et al. 2016) and Lago Yehuin (Markgraf 1983), the abrupt decrease in total pollen concentration values has been interpreted as a result of the instantaneous deposition of the thick H1 tephra layer. This process probably resulted in increased mineralization of organic matter by microbial activity and an alteration in the trophic and aquatic environment. Thus, the development of new habitats suitable for pioneer plants was manifested by an increase in grasses, herbs, shrubs and sedges. In La Correntina peat bog (Musotto et al. 2016), from 6300 cal yr BP and after an interval of ca. 1500 yr, *Nothofagus* pollen increased, reaching pre-eruptive frequencies, and at 5300 cal yr BP, the forest adopted closed-canopy characteristics. In Terra Australis peat bog a lower impact on the arboreal vegetation was recorded, with its pre-eruptive relative values recovered at 7500 cal yr BP and the definitive physiognomy of the closed-canopy *Nothofagus* forest established at around 6500 cal yr BP (Musotto et al. 2016, 2017). After ca. 8600 cal yr BP, the extent of grass, herb and shrub communities indicated conditions of decreased precipitation of < 400 mm annually compared to present-day vegetation. In western Tierra del Fuego, Mansilla et al. (2016) identified a significant reduction in forest cover at Punta Yartou, along with high charcoal values suggesting an intense arid phase after ca. 8080 cal yr BP.

Archaeobotanical data from Binushmuka I (Layer S, 8090 cal yr BP) and Imiwaia I (Layer S, ca. 8420–8660 cal yr BP), corresponding to periods prior to the H1 eruption, show assemblages dominated by *Berberis* sp. and *Embothrium coccineum*. These taxa are characteristic of open forest or shrubland environments and are recognized as pioneer species that tolerate high light conditions and facilitate the subsequent establishment of *Nothofagus* species. The dominance of these shrubs and small trees indicates that, prior to the H1 tephra deposition, the northern coast of the Beagle Channel was covered by relatively open vegetation. In contrast, archaeobotanical assemblages dated to the Middle Holocene (Binushmuka I Layer C and Imiwaia I Layer M, ca. 6720–6520 cal yr BP) reflect the expansion of closed-canopy *Nothofagus* forests—deciduous and mixed—as evidenced by the increased abundance and ubiquity of *Nothofagus pumilio* and *N. antarctica/betuloides* (Franch Bach et al. 2025).

2.3.4 Neoglaciations

Holocene neoglaciations or glaciations have been recognized in some sectors of the Patagonian Andes (Mercer 1968, 1982; Rabassa et al. 1984). Although their chronology remains imprecise, dendrochronological (Villalba 1990) and geomorphological (Rabassa et al. 1992) studies indicated the existence of at least five fluctuations during the Holocene related to climatic oscillations.

In Tierra del Fuego there is no evidence of significant neoglacial advances prior to the Little Ice Age (LIA). Menounos et al. (2013) reported geological evidence of one or more glacial advances in the hanging valleys of the Fuegian Andes sometime between 7960 and 7340 cal yr BP and 5290–5050 cal yr BP, reaching limits only tens of meters beyond the maximum extent of the LIA.

Pollen records from sites in the Beagle Channel and Isla de los Estados do not provide clear evidence of Holocene glacial fluctuations. However, intervals of reduced *Nothofagus* pollen input between 6380 and 5000 cal yr BP at the Harberton peat bog (Heusser 1989), between 6310 and 4510 cal yr BP at the Ushuaia 2 peat bog (Heusser 1998), and declines in pollen concentration around 6800 cal yr BP and between 5000 and 4500 cal yr BP on Isla de los Estados (Ponce et al. 2011) appear to correspond to colder climatic episodes (Coronato et al. 2022). Maximum tree concentration values, recorded after approximately 4900 and 4500 cal yr BP in southern Tierra del Fuego (Beagle Channel area) (Heusser 1989, 1998) and after around 4000 cal yr BP on Isla de los Estados (Ponce et al. 2011), indicate high effective humidity associated with the intensification of SWW at these high latitudes (Ponce et al. 2011; Björck et al. 2012). On the other hand, low sea surface temperature (SST) was reported in the Beagle Channel. According to Obelic et al. (1998) the SST was 1.5 °C below the present temperature between 3800 and 3500 cal yr BP, based on the isotopic composition ($\delta^{18}O$) in *Mytilus edulis* shells from archaeological sites from the Beagle Channel.

Pollen records from Isla de los Estados (Ponce et al. 2011; Björck et al. 2012) indicate that very cold and wet conditions ended with an abrupt decline in *Nothofagus* pollen around 3000 cal yr BP, suggesting a shift to warmer and drier conditions, potentially driven by a weakening of westerly winds. This transition is also evident in pollen records from the Lago Fagnano area (Musotto et al. 2016), where closed-canopy forests were replaced by open *Nothofagus* forests, accompanied by the expansion of mesic grasslands (Poaceae, Caltha and Cyperaceae), dwarf shrub heath (*Empetrum rubrum*) and cushion heath (Azorella and *Drapetes muscosus*). A concurrent climatic improvement was observed in the Beagle Channel, marked by an increase in sea surface temperatures (SST) shortly before 3000 cal yr BP (Obelic et al. 1998).

2.3.5 Medieval Climate Anomaly and Little Ice Age

Two climate anomaly events occurred during the last millennium, the so-called Medieval Warm Period -also known as the Medieval Climate Anomaly (MCA, 1100–800 cal yr BP)- and the Little Ice Age (LIA, 600–100 cal yr BP) (Moreno et al. 2014). The MCA date range was based on extratropical tree-ring data from the northern hemisphere (Esper et al. 2002). According to Broecker (2001), evidence for a global MCA is circumstantial and paleoclimatic data are relatively sparse in the southern hemisphere. Mentions of the MCA in the pollen records from Tierra del Fuego are limited. Mauquoy et al. (2004) mentioned a period of low local water table in the Andorra Valley peat bog (in an inland valley of the Fuegian Andes) between 1000 and 860 cal yr BP (AD 1020–960). Waldmann et al. (2010) identified the MCA in a sedimentary core from Lago Fagnano based on intervals of low iron content, which they interpreted as indicative of reduced precipitation and concurrent glacial retreat.

The LIA comprises a period when glaciers in mountainous areas of the world were generally more extensive. In the Fuegian Andes region, this event was one of the most widespread neoglacial advances (Menounos et al. 2013) and was identified in pollen records from Tierra del Fuego (Heusser 1989; Borromei et al. 2010, 2016; Musotto et al. 2016) and Isla de los Estados (Ponce et al. 2017). At La Correntina peatland, located at the southeastern end of Lago Fagnano, the reductions in the frequency and concentration of arboreal taxa together with an increase in pollen of *Empetrum rubrum* and Azorella around 400 cal yr BP were indicative of a reduction of *Nothofagus* forest and a drier peatland surface (Musotto et al. 2016). In the inland Andean valleys, in the vicinity of the Beagle Channel, *Nothofagus* forest declined during the last 1000 cal yr BP, reaching a minimum between ca. 680 and 300 cal yr BP (Borromei et al. 2010, 2016). In addition, a period of cooler and/or wetter conditions between ca. 920 and 850 cal yr BP and later in ca. 150–20 cal yr BP have also been recorded in the Andorra Valley peatland (Mauquoy et al. 2004). On Isla de los Estados, density in forest communities reached a minimum between ~500 and ~50 cal yr BP (Ponce et al. 2017). The decrease in *Nothofagus* pollen concentration values is strongly consistent with the period of colder surface waters in the Beagle Channel, at least during the last 4000 years (Ponce et al. 2017).

2.4 Archaeological Demography

To evaluate relative variations in population parameters, we analyzed potential changes throughout the Holocene using radiocarbon dates from the north coast of the Beagle Channel, the southeastern margin of Península Mitre and Isla de los Estados. This analysis included radiocarbon information obtained from charcoal and guanaco (*Lama guanicoe*) bones, as reported in previous studies (Zangrando et al. 2009; Horwitz and Weissel 2011; Orquera and Piana 2020).

In recent decades, though not without criticism, archaeologists have frequently resorted to Summed Probability Distributions (SPD) as a proxy for discussing population trends. While peaks and valleys in SPD may reflect changes in date intensity over time (and thus be interpreted as demographic "ups" or "downs"), they may also be a consequence of the changing steepness of the calibration curve, the associated measurement errors and/or simply a product of sample size (Williams 2012; Shennan et al. 2013; Crema et al. 2017). In response to these challenges, several studies have developed new techniques to address some of these problems (Shennan et al. 2013; Timpson et al. 2014). In particular, they have offered new approaches to the problem of discerning genuine fluctuations in ^{14}C date density as opposed to statistical artifacts arising from sampling errors, the calibration process or from the effect of taphonomic histories. In our analysis, we followed the calibration and probability summation procedure proposed by Crema and Bevan (2021). Only dates with standard deviation (SD) less than 150 years were considered. The dates were calibrated with the SHCal20 (Hogg et al. 2020) curve with the rcarbon package for R software (Crema and Bevan 2021). To reduce the potential effect of bias from summing samples from the same site that are close in time, the binning method proposed by Shennan et al. (2013) was used by applying a parameter h of 50 years. A number of potential biases, including taphonomic loss, sampling error and the shape of the calibration curve, can affect the shape of empirical SPD. One way to address these biases is to evaluate SPD against theoretical expectations using a hypothesis-testing framework. We employ the Monte-Carlo method proposed by Shennan et al. (2013) and Timpson et al. (2014). We therefore fit the unified empirical SPD for the three sectors of the southern coast of Tierra del Fuego to the null exponentially growing population model, which was fitted using the "exponential" option implemented in the model Test function of the rcarbon package for R (Crema and Bevan 2021).

Figure 2.3 presents the SPD obtained and Fig. 2.4 integrates this information with major paleoclimatic events to facilitate visual comparison with paleodemographic trends. In the Beagle Channel (183 radiocarbon dates from 84 archaeological sites), population continuity is observed throughout most of the Holocene. Between 8600 and 7500 cal yr BP, the frequency of occupations remains low, which may be due to a lower visibility of the coastal archaeological record (Zangrando 2018; Zangrando et al. 2021a). From 7500 cal yr BP there is a significant increase in occupations that lasts until 6300 cal yr BP and is characterized by two distinct pulses: one around 7300 cal yr BP and another around 6700 cal yr BP. After 6300 cal yr BP a demographic trend is projected that remains very low until 2000 cal yr BP. From this time onwards, a trend of population increase is observed, with a maximum peak at 500 cal yr BP. From this date onwards, the paleodemographic curve decreases toward contact times.

On the southeastern coast of Península Mitre, where the radiocarbon sample remains limited (25 radiocarbon dates from 14 archaeological sites), population discontinuity is observed for most of the Holocene until about 2700 cal yr BP; the occupations recorded prior to that date occur in the sequence of the same site (BVS11) between 7000 and 4700 cal yr BP. From 2700 cal yr BP there is a relative continuity at the population level until around 500 cal yr BP, when there is an

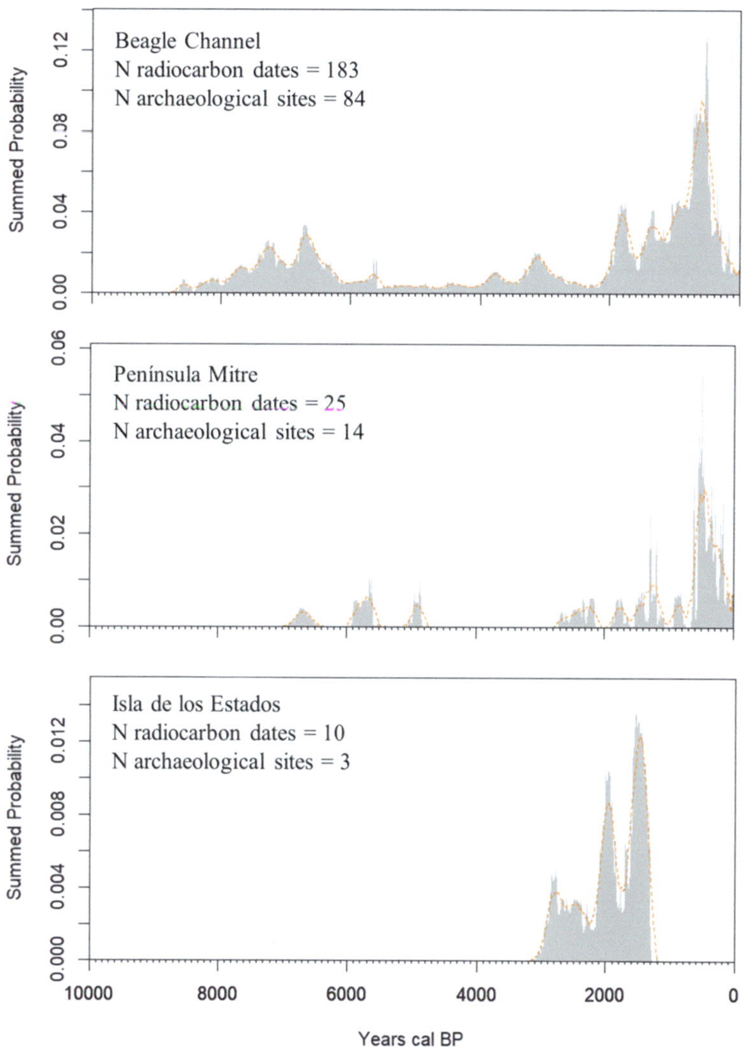

Fig. 2.3 Summed probability distributions (SPD) from the Beagle Channel, Península Mitre and Isla de los Estados

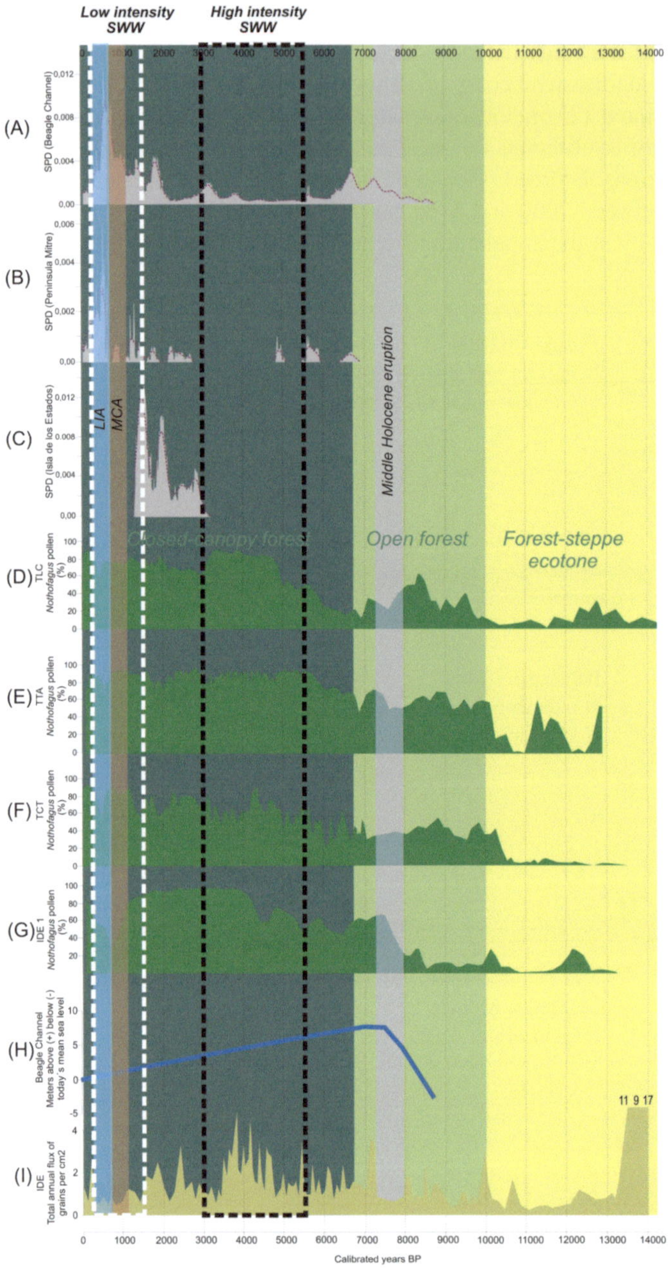

Fig. 2.4 Summed probability distributions (SPD) from **a** Beagle Channel, **b** Península Mitre, and **c** Isla de los Estados aligned with major paleoclimatic events, **d** La Correntina peat bog *Nothofagus* sp. pollen percentage (Musotto et al. 2016), **e** Terra Australis peat bog *Nothofagus* sp. pollen percentage (Musotto et al. 2017), **f** Cañadón del Toro peat bog *Nothofagus* sp. pollen percentage (Borromei et al. 2016), **g** IDE1 peat bog *Nothofagus* sp. pollen percentage (Ponce et al. 2011), **h** Beagle Channel Holocene sea-level curve (Björck et al. 2021), and **i** Isla de los Estados total annual flux of grains (Björck et al. 2012)

increase in the number of occupations. The trend described after this date is similar to that recorded in the Beagle Channel. Isla de los Estados only records occupation between 3000 and 1700 cal yr BP, with three more marked moments in 2800, 2000 and 1500 cal yr BP. It should be noted that the temporal record of occupations for the island is reduced to 10 radiocarbon dates from 3 archaeological sites.

One striking aspect is the greater population continuity observed in the Beagle Channel compared to the far eastern part of Tierra del Fuego. As noted earlier, this difference may be partly attributed to variations in sample sizes. Nonetheless, it is noteworthy that population increases are recorded during the same period in the Late Holocene for both the Beagle Channel and Península Mitre. While we cannot entirely rule out the possibility that the discontinuities and short occupation spans in the eastern sector result from limited research intensity or fewer radiocarbon dates, additional archaeological evidence—discussed in the following sections—supports the interpretation of spatial variation in occupational patterns.

The analysis of the unified empirical SPD for the three sectors of the southern coast of Tierra del Fuego ($N = 218$), adjusted to the null population model with exponential growth, indicates a significant demographic increase between 7500 and 6300 cal yr BP (Fig. 2.5). After this period, and up to 4800 cal yr BP, the population size tends to decrease. The lowest population density is recorded between 4800 and 2000 cal yr BP, a period in which the radiocarbon dates (black line) is observed to cross the lower limit of the null exponential fit model with a 95% confidence interval (gray-shaded area). During this phase, five negative deviations (blue shaded areas) are identified, the most significant occurring between 3500 and 3300 cal yr BP, which may suggest a population collapse. An interesting aspect to consider is that this period corresponds to the peak intensity of the SWW (Fig. 2.4) (Björck et al. 2012). After 2000 cal yr BP, there is a notable increase in population density, with similar patterns recorded across all sectors. Two significant demographic expansions are observed: one at approximately 1800 cal yr BP and another around 500 cal yr BP, the latter representing the largest population increase in the entire sequence. Following this peak, a decline in population is recorded, continuing until the period of contact.

2.5 Archaeological Landscape

2.5.1 Colonization

The prevailing hypothesis suggests that the initial occupation of the insular region of Fuego-Patagonia occurred later than in the interior of continental Patagonia. This process is believed to have involved seaworthy adaptation, followed by the dispersal of these groups across the archipelago (Orquera and Piana 1987, 2005, 2006; Legoupil and Fontugne 1997; Ocampo and Rivas 2004; Prieto et al. 2013). While there is no consensus among these authors regarding the specific areas from which this dispersal originated, it is proposed that this process began around 7300 cal yr BP.

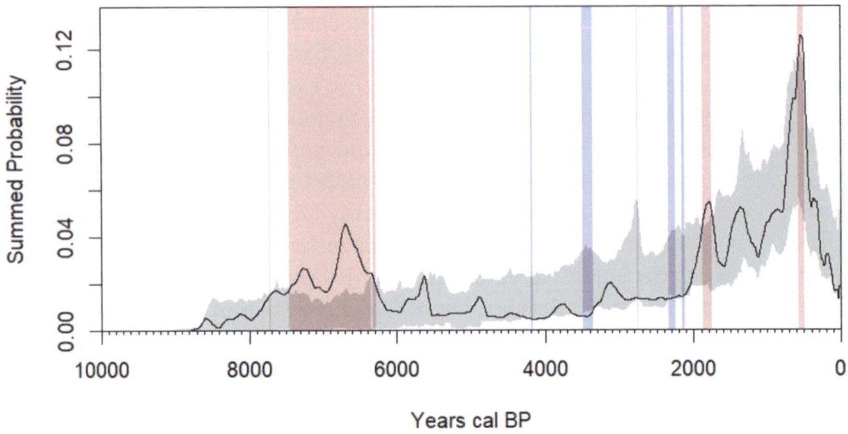

Fig. 2.5 Summed probability distribution (SPD) of radiocarbon dates from the south coast of Tierra del Fuego, fitted to an exponential growth model. The gray-shaded area represents the critical envelope encompassing the central 95% of the simulated SPD, while the red and blue regions indicate statistically significant positive and negative deviations, respectively

This timeline is based not only on available chronological evidence but also on the emergence of zooarchaeological assemblages dominated by marine species and the appearance of specialized technologies for exploiting these resources, as observed in shell middens. According to some authors, the initial occupation of the archipelago coincided with two distinct natural processes that likely influenced adaptation to the coastal environment and subsequent dispersal. One was gradual: the expansion of the *Nothofagus* forest (Orquera and Piana 1987, 2005, 2006); the other, catastrophic: the eruption of the Hudson volcano (H1) (Prieto et al. 2013).

Orquera and Piana (1987, 2005) propose that a transition from terrestrial to canoe-based mobility among HGF populations occurred along the western shores of the Strait of Magellan and the Otway and Skyring Sounds. From this area, a subsequent dispersal to the archipelago is believed to have taken place. One of the key factors supporting this hypothesis is the role of the *Nothofagus* forest expansion, which was critical for the development of navigational technology and the specialized exploitation of marine resources. According to the pollen studies initially referenced by the authors (Heusser 1989, 1998), the development of this ecosystem occurred earlier north of the Strait of Magellan than in the Beagle Channel. However, two factors should be considered in relation to this argument. First, recent studies based on pollen profiles indicate that closed-canopy forests were established from latitude 45°S (northern Santa Cruz) to 54°S (Beagle Channel) during the period 11,000–12,500 cal yr BP and have persisted since then, albeit with variations on millennial and centennial timescales (Moreno et al. 2018; McCulloch et al. 2020). Among the most notable environmental changes is a prolonged arid phase (Fig. 2.4), lasting several millennia from approximately 10,100–6800 cal yr BP (McCulloch et al. 2020), which broadly encompasses the Extended Warm-Arid Anomaly (Moreno et al. 2018). This period is marked by a significant retraction of forests across the entire latitudinal range. As a result, it is difficult to support substantial geographical differences in the availability of raw materials required for the production of navigation technology. Additionally, recent archaeobotanical evidence from coastal sites in the Beagle Channel area, demonstrates the use of shrub and tree taxa from at least the Early Holocene (Franch Bach 2024). Second, following the arid period, there was a restoration of closed *Nothofagus* forest from ca. 6800 cal yr BP (McCulloch et al. 2020). This record challenges the proposed link between the expansion of *Nothofagus* forests and the dispersal of canoe populations in the region (Orquera and Piana 1987), since the notable population growth recorded in the southern region of Tierra del Fuego between 7500 and 6300 cal yr BP (Fig. 2.4) occurred before the expansion of closed *Nothofagus* forest in the Beagle Channel.

Orquera and Piana (1999, 2009) recognized the presence of coastal occupations prior to the emergence of specialized littoral adaptations, but they attributed these settlements to terrestrial hunters who occasionally visited coastal areas. Their argument was based on techno-typological criteria, particularly differences in the degree of bifaciality in lithic assemblages and the appearance of bone technology after 7300 cal yr BP (Orquera and Piana 1999, 2009; Zangrando 2009:103–104; Piana et al. 2012; Orquera 2022). However, recent studies have highlighted the need to reassess this perspective, suggesting that the visibility and preservation of the archaeological record may vary across different periods and be influenced by the timing and nature of the marine transgression. This process likely had significant effects on both landscape transformation and human behavior, which must be considered when interpreting the record (Zangrando 2022; Zangrando et al. 2021a).

A more recent proposal, put forward by Prieto, Stern and Estévez (2013), supports the onset of coastal adaptation in the Brunswick Peninsula and Otway Sound region during the Middle Holocene. Additionally, these authors suggest that the eruption of the Hudson volcano (H1) -assuming that this event caused widespread devastation of the terrestrial ecosystem and possibly led to the annihilation of herbivores such as the guanaco- may have driven humans to specialize in exploiting marine resources, which remained unaffected by the eruption. Based on these arguments, they propose a recolonization of Tierra del Fuego by canoe-using groups, hypothesizing that the eruption likely led to the extermination of terrestrial hunter populations on the island. A key expectation of this model is that the archaeological sequence along the southern coast of Tierra del Fuego would exhibit a disruptive pattern, marked by regional-scale hiatuses corresponding to the presumed processes of transition, population replacement and/or dispersal between terrestrial and marine hunter-gatherers. However, as shown in Figs. 2.4 and 2.5, no temporal hiatuses or disruptions are recorded in coastal occupations of the Beagle Channel between 8600 and 6000 cal yr BP on a regional scale, a period that overlaps with event H1. In fact, this period shows a recorded population increase. Furthermore, the impact of the H1 event on the terrestrial ecosystem of the region remains unclear, highlighting the need for additional paleoecological studies to determine whether the eruption had catastrophic effects. While pollen records indicate a sharp decline in total pollen concentration values, this may be attributed to the instantaneous deposition of a thick layer of H1 tephra. Additionally, the accumulation of ash could have increased the mineralization of organic matter through microbial activity, potentially altering trophic and aquatic environments. These changes might have led to the development of new habitats, including increases in grasses, herbs, shrubs and sedges. Finally, it is important to note that although zooarchaeological remains of guanacos are relatively sparse immediately after the H1 eruption, this does not support the idea of species extinction on the island (Fernández et al. 2020).

As we have argued elsewhere (Zangrando 2018), these observations lead us to suggest that the colonization of the insular areas of Tierra del Fuego may have occurred before the Mid-Holocene. The colonization of a new territory typically involves a period of low population density, resulting in exceptionally poor preservation and low archaeological visibility (Borrero 1994–95; Steele and Politis 2009; Prates et al. 2020). This issue is even more pronounced in coastal environments, where the analysis must account for the dynamic nature of the landscape, which is constantly transformed by marine transgression, geomorphological changes, and their impact on the visibility and preservation of archaeological evidence (Zangrando 2022). Consequently, it is reasonable to assume that the earliest documented archaeological records in a region appear sometime after the initial human dispersal in that territory.

From this perspective, it is challenging to support the idea that this process in the archipelago south of Tierra del Fuego occurred around 7300 cal yr BP, given the demographic continuity and population growth recorded in the region, with a significant rise between 7500 and 6300 cal yr BP. This population trajectory is inconsistent with expectations for a colonization period. In fact, there is evidence of earlier human activity in the Beagle Channel. For instance, at Gable Island, a stone ball was recovered in a detached block from a peat stratigraphic profile. Radiocarbon dating of wood samples from two distinct peat levels—one above and one below the artifact—yielded calibrated ages of 9800 and 10,800 cal yr BP. According to one of these studies, the stratigraphic position of the stone ball corresponds to the level of the earliest dated sample (Salemme et al. 1995; Rabassa et al. 2000).

We have recently initiated a search for archaeological deposits in the region, focusing on geomorphological units that may contain evidence of early coastal occupations. Our aim is to explore the coastal space and interpret the archaeological record from a perspective that considers different conditions of formation, visibility and preservation than those typically associated with shell middens (Tivoli et al. 2022; Zangrando et al. 2021a). The emphasis on exploring coastal areas for early sites outside of mound formations stems from viewing the space affected by marine transgression as a taphonomically active zone (Borrero 2014), also referred to as landscape taphonomy (Bailey 2014). This approach calls for moving away from categories such as "maritime" ("canoe-using") and "terrestrial" to characterize these populations, as such labels oversimplify the issue and overlook the cultural dynamics between populations with different subsistence strategies–dynamics that are to be expected during a colonization process.

2.5.2 *Shell Middens and Coastal Archaeology*

From the Middle Holocene onwards, shell middens were the most prominent archaeological feature along the southern coast of Tierra del Fuego (Fig. 2.6). Although other types of archaeological contexts exist in the region -burials, lithic scatters or isolated finds- their visibility is much lower than that of the shell middens. These non-shell midden sites are often obscured by the development of regional vegetation. Rock shelters are relatively rare, and their use has been associated with burial or artistic practices (Piana et al. 2006; Gallardo et al. 2023). While these factors should not be overlooked in analyzing the regional occupation pattern, shell middens remain the most abundant and systematically studied archaeological features, providing the richest source of information about past human behavior in southern Tierra del Fuego (Orquera and Piana 1999). Moreover, the significance of these deposits for archaeological research in coastal regions is well established, as their formation reflects behavioral and evolutionary processes operating across multiple scales (Bailey 1983). In the case of our study, it is crucial to consider:

(1) Both ethnographic and archaeological evidence from southern Tierra del Fuego suggest that shell middens were typically formed in close proximity to habitation areas, or even encircling them. As a result, these shell matrices preserve a significant diversity of features and material remains, which reflect various behavioral aspects of HGF groups. Among the preserved materials are traces of hearths, lithic and bone tools, personal ornaments, pigments and other cultural artifacts.

Fig. 2.6 Shell middens at Cambaceres Bay, Beagle Channel (Photo: JS Eidshaug)

(2) Archaeological excavations in the region have also shown that the formation of shell middens resulted from repeated occupations in the same place (Orquera et al. 1977; Orquera and Piana 1986–87, 1996, 1999; Yesner 1990; Estévez and Vila 1995, 2006, 2007; Piana et al. 2004; Álvarez et al. 2009; Zangrando 2010; Zangrando et al. 2016). These reoccupations are indicated both by continuous accumulation of shells over short timescales and by the reuse of the same site after extended periods—ranging from centuries or millennia. As such, shell midden formation has been defined as the result of long-term practices in which the accumulation of shell deposits promoted their subsequent reuse (Piana and Orquera 2010).
(3) Shell middens can form dense clusters along coasts, to such an extent that they represent a prominent anthropogenic feature in the coastal environment. These clusters may have more than a hundred interconnected structures and are functionally distinctive nodes in the mobility patterns of HGF societies. This conspicuous visibility in the landscape provides advantages for exploring the cultural geography of a region (Binford 1982; Haas and Kuhn 2019).

Orquera and Piana (1989–90) identified various shell structures along the north coast of the Beagle Channel, with dome and ring-shaped formations being the most common. A significant proportion of these are ring-shaped, with a depressed center approximately three to four meters in diameter. Ethnographic and archaeological evidence suggests that this concavity marks the location of the dwelling. Regardless of their form, shell mounds generally reach heights of about one meter, although higher dimensions have been recorded, as well as deposits that are only a few centimeters thick.

Most of the shell structures on the north coast of the Beagle Channel are situated below 10 m above sea level, and nearly 90% are found within 0 to 300 m of the current shoreline (Barceló et al. 2002; Orquera and Piana 2009). Several localities within the region exhibit very high site densities. For instance, in and around Bahía Cambaceres, 1237 shell structures were documented in an area of just 4 km^2 (Bjerck et al. 2016). However, such high site densities are not necessarily uniform across the entire region. A recent study by Eidshaug et al. (2024) found that the highest frequency of ring-shaped shell structures is concentrated in the Beagle and Murray channels. However, these dense clusters of ring structures are not observed in areas of the archipelago that were actively occupied by the HGF during the Late Holocene, such as the Cape Horn Archipelago, Hardy Peninsula in Hoste Island, Nassau Bay, and Picton, Lennox, and Nuevo Islands. This suggests that the observed regional variability in the frequency of shell middens may be more closely linked to differences in the intensity of island occupation (Eidshaug et al. 2024). In contrast, dome-shaped shell structures were found to be more widely distributed throughout the Fuegian Archipelago. This is particularly true along the southern coast of Isla Grande, where mounds are the predominant feature from Moat to the eastern end of the Fuegian region. It is also worth noting that dome-shaped structures are frequent in areas with high densities of ring-shaped shells. For example, Bjerck et al. (2016) identified 432 domes among the 1237 shell midden sites recorded during the Cambaceres Bay surveys.

The identification of sites located beyond the narrow coastal margin is truly exceptional. While it is possible that the mountain valleys may have been sparsely and sporadically occupied, these areas remain underexplored and warrant more systematic and intensive archaeological investigation. The use of interior valleys may have been more closely tied to the procurement of specific resources or to movement and circulation. However, according to ethnographic evidence, residential occupations in these areas cannot be ruled out, at least in later times of the Holocene. While the coasts would have provided significant advantages for human subsistence in terms of resource abundance, diversity and stability, research from inland lakes in Tierra del Fuego indicates that the sub-Antarctic forest was also actively occupied by HGF populations (Mansur and Piqué 2009; De Angelis 2015). Understanding the nature and extent of the use of the Fuegian Andes valleys, particularly along the southern coast of Tierra del Fuego during the Holocene, remains a significant knowledge gap in the region.

2.6 Technological Variations

In archaeological assemblages from the southern coast of Tierra del Fuego, lithic tools are basically composed of three categories: long retouched edges, scrapers and flakes with natural edges that present a significant internal variation. Other typological groups, such as weapon points or polished or pecked tools, appear less frequently represented in the assemblages (Álvarez 2009; Zangrando et al. 2009; Horwitz and Weissel 2011). Although this general trend is evident throughout the archaeological sequence of the region (Orquera and Piana 1999), temporal variations in the morphological diversity of lithic tool assemblages have been documented in both the Beagle Channel (Orquera and Piana 1999; Álvarez 2009; Hernandez Herrero 2024) and Península Mitre (Zangrando et al. 2009; Vázquez et al. 2011). Similar patterns have also been observed in other parts of the archipelago (e.g., Legoupil 2003; Orquera et al. 2011; San Román et al. 2016). Álvarez (2009) identifies two key turning points in lithic technological variations in the Beagle Channel. The first, dated to approximately 4400 cal yr BP, is marked by the emergence of a new design in hunting tools. These tools are characterized by large lanceolate points (Orquera et al. 1977, 2011; Legoupil 2003). Notably, these points are distributed across the entire Patagonian archipelago, having been documented in the northernmost regions, such as the Reloncaví Sound (Gaete and Navarro 2004) and the Chonos Archipelago (Reyes 2021) (5800–4600 cal yr BP), as well as in the Beagle Channel (4400 cal yr BP) (Orquera et al. 1977; Piana et al. 2004). Their presence has also been confirmed along the Strait of Magellan, on the eastern coast of the Brunswick Peninsula (Rey Felipe, Km 44), Offing Island (Legoupil 2023) and the Última Esperanza Sound (Orquera et al. 2011). In all instances where these points have been dated, the chronology consistently aligns with the period 5800–4400 cal yr BP.

The second inflection point, emerging around 2700 cal yr BP, is characterized by increased variability in the composition of lithic tool assemblages. In particular, a morphological diversification is observed in the weapon points, which led to a greater incidence of bifacial knapping (Piana and Orquera 2007; Alvarez 2009). In contrast, only a few variants are observed in the processing tools. Although the designs of retouched and natural long edges remain consistent (Piana and Orquera 2007; Álvarez 2009), morphological changes are documented, along with variations in manufacturing techniques and the use of end-scrapers, between 2700 and 1000 cal yr BP.

During this later period, Piana and Orquera (2007) note numerous detailed parallels between lithic weapon points from the Beagle Channel and specimens from hunter-gatherer contexts on the Atlantic coast, suggesting that these similarities may reflect possible interactions between these populations. Ratto (1994) proposes distinguishing among three technical systems: throwing spears, bows and arrows and handheld weapons. This researcher observed that the coexistence of these systems in southern Patagonia and Tierra del Fuego was primarily driven by functional considerations rather than cultural or chronological factors (Ratto 1994). Such functional variation during the Late Holocene has been documented in assemblages from both Península Mitre (Ratto 1991) and the Beagle Channel (Orquera and Piana 1999; Álvarez 2009, 2011). Charlin et al. (2014), employing geometric morphometry analysis, identified a more homogeneous pattern of pedunculate point representation in the Beagle Channel compared to other regions of Tierra del Fuego. They observed common point designs in samples from the northern and southeastern parts of Isla Grande, whereas the Beagle Channel assemblages were less dominated by this form, which is more prevalent in the rest of Tierra del Fuego. This pattern suggests that projectile points from the northern regions and Península Mitre share greater morphological similarities than those from the Beagle Channel. Charlin et al. (2014) hypothesize that these differences could reflect variations in the degree of interaction among these island regions during the Late Holocene. For the contact period, Charlin et al. (2016) observed morphometric differences between glass arrowheads from the late nineteenth century and lithic weapon points collected by ethnographers, possibly from archaeological sites. They also suggest that these differences may result from functional factors.

The archaeological assemblages of southern Tierra del Fuego exhibit significant diversity in bone technology, predominantly represented by various forms of points and bevels (Chapman 1987; Scheinsohn 1997; Orquera and Piana 2009; Vidal (2011 [1985]); Horwitz and Weissel 2011; Tivoli 2013). After 2700 cal yr BP, changes in the basal shapes of harpoon points and/or the arrangement of their teeth are documented (Orquera and Piana 1999; Piana and Orquera 2007). Scheinsohn (2010) highlights that between 7500 and 1500 cal yr BP, considerable development occurred, marked by a certain degree of standardization within specific technical groups. This period also reflects the consistent use of four different raw materials—cetacean, camelid, bird and pinniped—which persisted throughout the entire regional occupational sequence. From 1500 cal yr BP onward, Scheinsohn (2010: 94) notes an increase in instrumental diversity, evident in the broader range of morphological types, raw materials and metric variability. Interestingly, this pattern of change aligns temporally with trends in lithic assemblages (Álvarez 2009). However, the relative frequency of bone tools compared to lithic tools declines after 1500 cal yr BP in both the Beagle Channel and the Península Mitre, a trend particularly notable in Valentín Bay (Zangrando et al. 2009). In this bay—one of the most intensively studied areas of Península Mitre—there is no evidence after 1500 cal yr BP of bird bones being used to manufacture awls (sensu Orquera and Piana 1986-87) or points (sensu Scheinsohn 1997, 2011). However, this absence does not necessarily indicate that artifacts made from bird bones were entirely lacking (Vázquez et al. 2011). Awls or points made from bird bones, which were used for basketry production during the contact period (Hyades and Deniker 2007 [1891]), are widely represented in archaeological assemblages across the region (Tivoli 2013). Their absence in the late assemblages from Valentín Bay is therefore striking. It is worth noting that seabird bones are abundant in the archaeological deposits at Valentín Bay, making preservation or sampling biases an unlikely explanation for this absence. One possible interpretation is that the activities performed in this area during later periods did not require the use of these tools (Scheinsohn 2011).

2.7 Social Networks

2.7.1 Lithic Raw Materials

Lithic raw material analysis offers valuable insights not only into technological change but also into labor organization throughout production sequences, technological choices influenced by material properties and functional needs, mobility patterns, circulation networks, social interactions, spatial organization and provisioning strategies related to lithic resource management (e.g., Renfrew 1977; Torrence 1989; Meltzer 1989; Nelson 1991; Álvarez et al. 2010; Alberti and Fernández 2015). A fundamental requirement for these studies is the identification of raw material source locations. In the Fuegian Archipelago, the exploitation of these resources primarily relied on secondary geological deposits (e.g., Álvarez 2009), which are one of the most common forms in which lithic raw materials are available (Andrefsky 1994). Discussions in this field have focused on two well-documented primary sources of raw materials, both of which have broad distributions throughout the region. One source is green obsidian, located near the Otway Sea and Riesco Island (e.g., Emperaire and Laming 1961; Ortiz-Troncoso 1973; Morello et al. 2015). The other, situated in the Chorrillo Miraflores Valley in northern Tierra del Fuego, consists of tuff and silicified tuff (Prieto et al. 2004; Borrazzo et al. 2019).

2.7.1.1 Green Obsidian

In Fuego-Patagonia, various obsidian sources show evidence of exploitation dating back to 10,000 years ago, with green obsidian[1] being the most widely distributed within the Magellanic-Fuegian archipelago (San Román and Pietro 2004; Stern 2004). While the exact location of the green obsidian source remains unknown, reviews suggest it is likely located near the Otway Sea and Riesco Island, northwest of the Strait of Magellan.

An analysis of 163 sites containing green obsidian reveals that over 90% of these sites are concentrated in this region. These sites are predominant coastal and typically feature nodules larger than 10 cm in length and over 50 g in weight (Morello et al. 2015). Among the identifiable blocks or nodules in the artifacts, 88% are in the form of baguette-shaped fragments or plates.

Green obsidian has a wide spatial distribution: along a north–south axis, it extends over 500 km, while in a west–east direction, its range reaches approximately 200 km (Morello et al. 2015). Beyond the archipelago, green obsidian has also been found in inland sites, such as the Cordillera Baguales and locations within the Palli Aike Volcanic Field (e.g., Cuevas Fell and Pali Aike).[2]

On the southern coast of Tierra del Fuego, green obsidian artifacts have been recorded at low frequencies in four Middle Holocene archaeological assemblages (Fig. 2.7). A lithic point was found in layer D of Túnel I (Trench XII - Level D6 – 7000 and 4800 cal yr BP). Obsidian artifacts were recovered in the Middle Holocene levels of Imiwaia I (layers K, L and M) with chronologies between 6700 and 6400 cal yr BP. Among them there is a bifacial tool made from a baguette-shaped nodule, measuring 97.22 mm in length and weighing 27.12 g. The type of nodule closely resembles those recorded near Otway Sea and Riesco Island. In addition, the assemblage includes a tool featuring a projecting "tooth," as well as a scraper. At Binushmuka I site, 143 obsidian artifacts were recovered, dated between 6800 and 6500 cal yr BP, including 139 flakes or flake fragments and four tools (Zangrando et al. 2018). The latter have similar morphological characteristics as the bifacial artefact with "tooth" found at Imiwaia I. Among the flakes or flake fragments, only 9% exceed 10 mm in length. Finally, at Heshkaia 60, where occupations are dated between 7100 and 6800 cal yr BP, 24 green obsidian artifacts were recovered, consisting primarily of micro-flakes (Zangrando et al. 2021b; Hernandez Herrero 2024).

---

[1] Early descriptions (Emperaire and Laming 1961; Ortiz-Troncoso 1973, 1975) suggest that this raw material exhibits a distinctive dark green to olive color in HGF contexts in Otway Sound. The distribution of the predominant green color is 87% homogeneous, although 10% of the artifacts show dark bands (Morello et al. 2015). Naked-eye identification of green obsidian has proven to be highly accurate (Stern 2000). Stern (2018) further notes that, in addition to its distinctive color compared to other Patagonian obsidian, green obsidian is also chemically unique.

[2] The earliest possible record of this raw material comes from the Palli Aike and Fell caves, which are assigned to Period III as defined by Junius Bird (1993) and correspond to a chronology between 9500 and 7400 cal yr BP. However, since other dates from this period suggest later ages (6740-6485 cal yr BP), which correspond to the Middle Holocene, Morello et al. (2015) argue that it

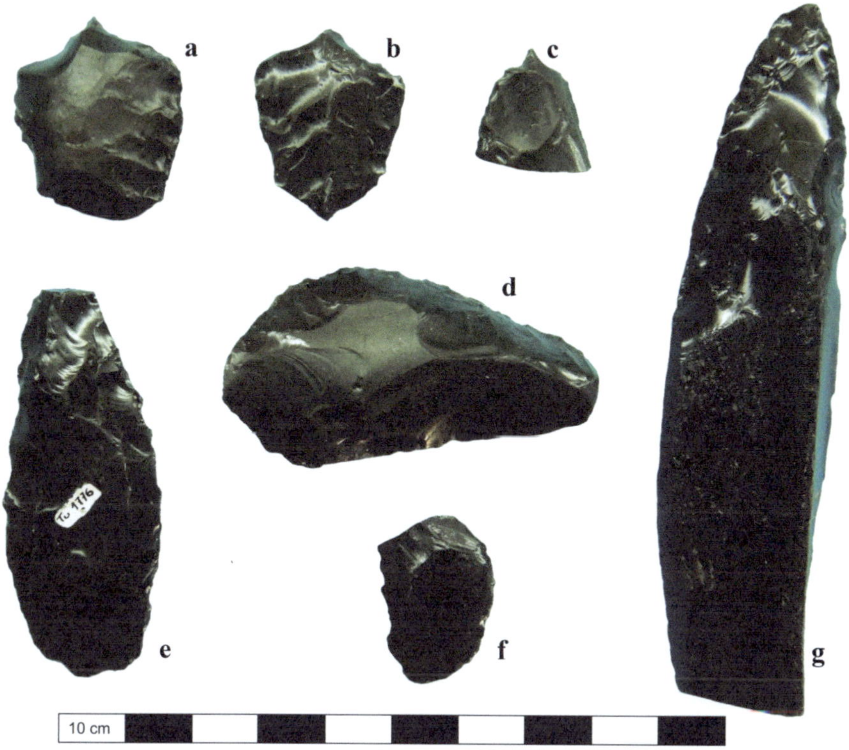

Fig. 2.7 Green obsidian tools from different sites in the Beagle Channel: **a–c** artifacts with projecting tooth and **d**, **f** scrapers from Imiwaia I and Binushmuka I, **e** lithic point from Túnel I and **g** bifacial on baguette-shaped nodule from Imiwaia I

The spatial and temporal distribution of green obsidian artifacts in the region has been analyzed from different perspectives, including geological (resource availability), biogeographical (barriers and circulation difficulties), ecological (terrestrial/marine subsistence) and cultural (interaction and exchange) aspects. The presence of green obsidian appears to be temporally discontinuous across the region (San Román and Pietro 2004). In the Beagle Channel, the record of this raw material is limited to a period between 7100 and 4800 cal yr BP. Only during this period is it possible to establish a connection between the Beagle Channel and areas adjacent to the Strait of Magellan. Recent studies carried out in Ocasión Sound, on the southwest coast of the Brecknock Peninsula (Tierra del Fuego), identified the Brecknock 1 site located on an esplanade at 10 m a.s.l., where lithic materials were identified, including mainly flakes and a green obsidian point on an erosive cut (San Román et al. 2023). This context is chronologically located between 7400 and 6300 cal yr BP and supports the existence of interaction networks between maritime hunter-gatherer groups in the

is uncertain whether this earlier chronology can be attributed to the presence of obsidian and recommend further chronological control.

Strait of Magellan and the archipelago south of Tierra del Fuego during the Middle Holocene. Evidence from this site also suggests the high performance of the navigational technologies employed by these groups and the importance of contacts and/or exchanges between the two geographical areas during this period (San Román et al. 2023).

There was a significant decrease in the use of obsidian until around 2500 cal yr BP, after which it reappears more frequently in archaeological sites throughout Magallanes area. Its presence continues into the contact period, when historical sources document its use. The regional availability of the resource, the connectivity of the landscape and the strategies of mobility, territoriality and exchange all influenced the distribution of artifacts. Morello et al. (2015) suggest that the discontinuity in the use of green obsidian is a key element for understanding human settlement patterns in the macroregion, particularly with regard to cultural and social dynamics. Those authors correlate this discontinuity with technological changes, noting that the sharp decline in green obsidian coincides with the emergence of large lanceolate points in the archipelago. On the other hand, Pallo (2016) argues that the long-distance transport of obsidian and other exotic raw materials was limited and flexible over time, reflecting the irregularity of social interaction networks (Borrero et al. 2011; Pallo and Borrero 2015).

2.7.1.2 Chorrillo Miraflores

Petrographic analyses of the raw materials from Miraflores have identified them as a tuff and a silicified tuff (Borrazzo et al. 2010, 2015). This raw material offers insights into alternative spatial vectors of human circulation and social interaction within Tierra del Fuego, distinct from those associated with green obsidian (Borrazzo et al. 2015, 2019; Pallo and Borrazzo 2016). The significance of Miraflores raw materials is underscored by the recovery of artifacts made from them at sites in the southern part of the island, on nearby smaller islands and along the coast of mainland Patagonia (Borrazzo et al. 2015). Their presence provides additional evidence of interactions among various HGF populations in Tierra del Fuego (Borrazzo et al. 2019).

The earliest use of Miraflores raw materials is dated to around 4000–3000 cal yr BP, with a more substantial presence in late Holocene assemblages (Pallo and Borrazzo 2016; Borrazzo et al. 2015, 2019). Miraflores tuff is a bright reddish rock with an eutaxitic texture, worked by grinding and used either for polishing wooden shafts or as a grinding stone. Silicified tuff, an isotropic rock, exhibits highly variable knapping quality depending on the frequency of impurities and the degree of silicification. Scrapers made from silicified tuff, one of the most curated and morphologically standardized tool types in Tierra del Fuego, are particularly common (Pallo and Borrazzo 2016). Considering the linear distance from the Miraflores source, located in the eponymous valley in the northern steppe (Fig. 2.1), Borrazzo et al. (2019) observed that cores are virtually absent in assemblages more than 55 km from the source. This pattern likely reflects a distance threshold beyond which the transport of complete nodules was no longer practical. However, the archaeological assemblage with the highest frequency of silicified Miraflores artifacts is found at the Kami 1 site, approximately 200 km from the source (De Angelis 2014, 2015). The southernmost occurrence of silicified tuff is recorded at Bahía Valentín, on the southeastern coast of Tierra del Fuego (Pallo and Borrazzo 2016).

As noted above, scrapers are the most common tool type among the silicified tuff artifacts and so far the only one identified on the southern coast of Tierra del Fuego from knapping debris. Borrazzo et al. (2019) found no correlation between the size of these tools and the distance to the Miraflores source, suggesting that the broad mobility ranges of human populations may have facilitated their widespread distribution. The authors also highlight the coexistence of these tools with other distinctive technological features, such as the bipolar exploitation of quartz pebbles, across much of Tierra del Fuego. This coexistence may reflect the large-scale circulation of Miraflores raw material during the late Holocene, a hypothesis further supported by ethnographic accounts (Chapman 1986; Gusinde 1990 [1937]; Bridges 2000 [1947]). Spatial analyses reveal that the circulation dynamics of Miraflores materials varied between northern and southern Tierra del Fuego. These differences may reflect diverse methods of raw material acquisition at Miraflores (direct or indirect) by populations in these areas, as well as differing degrees of reliance on social interactions (Pallo and Borrazzo 2016). In the northern region, multiple routes were likely utilized, facilitated by the permeability of the Fuegian steppe and key landscape features such as river valleys and the coasts of the Strait of Magellan, as well as Inútil and San Sebastián bays. Conversely, in southern Tierra del Fuego, connections to the Chorrillo Miraflores source appear limited to optimal routes leading to sites at the eastern end of the island.

Although the presence of Miraflores raw materials in the Beagle Channel has not yet been documented, the potential involvement of canoe populations in their distribution cannot be ruled out. Notably, the circulation of silicified tuff across the Strait of Magellan has been indirectly recorded (Langlais and Huidobro 2023), raising the possibility that maritime groups may have contributed to its wider dispersal.

2.7.2 *Portable Art*

Art can be considered a mechanism of construction of social-demographic dynamics both at inter-group and intra-group scales, since it enables communicative, expressive, aesthetic and performative practices (Conkey 1980; Jochim 1983; Robb 2017; Nowell 2017; Fiore 2020). As such, it provides a valuable lens through which to explore social interaction (Gamble 1982), resilience (Fiore 2014; Romero Villanueva et al. 2024) and niche construction (Acevedo 2024) in past human populations. From this perspective, portable art has been studied as an indicator of mobility and interaction among HGF populations in the Fuegian Archipelago. These studies have examined the continuity and discontinuity of decorated bone artifacts as part of visual communication networks (Fiore 2006, 2011), with a particular focus on two types of objects: harpoon points and beads.

Decorated harpoon points are predominantly associated with deposits dating from 7300 to 4500 cal yr BP (Fig. 2.8). These artifacts show a diverse range of engraved designs mostly created with sinuous and straight lines, figures, dashes and few cases of dots, which suggests that individual innovation played a significant role in their decoration (Fiore 2011). However, similar motifs are found across distant areas of the archipelago, such as the Otway Sound and the Beagle Channel (Fiore 2006). Similarities in the harpoon types chosen for decoration, in motif designs and engraving techniques occurred and decreased at similar times in these different regions, thus providing evidence of effective social interaction throughout several areas of the Fuegian Archipelago (Fiore 2006, 2011).

In contrast, decorated beads made with bird bones display greater continuity throughout the archaeological sequence. These artifacts are characterized by simple, highly standardized designs based on sets of dashes and sets of lines (rings or spirals), with low variability in their decorative repertoire. Unlike harpoon points, individual innovation appears to have played a limited role in bead decoration, with transmission mechanisms like teaching, learning and imitation possibly driving the production of these personal ornaments consistently over 6400 years (Fiore 2006). Beads with these designs are widely distributed across Fuego-Patagonia, appearing in the Beagle Channel (Orquera and Piana 1999), northern Tierra del Fuego (Massone 1988; Massone et al. 1993), Otway Sound (Legoupil 2003), Brunswick Peninsula (Ortiz-Troncoso 1979) and the Palli Aike Volcanic Field (Bird 1993). These space and time continuities in bead production and decoration are compatible with the construction of a collective, broad-scale and inter-generational social identity, in which the presentation of the "personal self" via these ornaments would have encouraged an active way of engagement within cohabiting groups and between spatially distant groups (Fiore 2020).

Thus, we suggest that the distribution of decorated harpoons (Fiore 2006), restricted to the Middle Holocene, and of decorated beads (Fiore 2011), spanning along the Middle and Late Holocene, reflect social connections facilitated by visual media. These media may have served as mechanisms of reciprocity, helping to buffer stress and mitigate risks (Sahlins 1977; Gamble 1982; Jochim 1983).

2.8 Genetic Diversity of Hunter-Gatherers-Fishers

The study of ancient genomes is significantly enhancing our understanding of past human interactions in the Americas (Llamas et al. 2016; Posth et al. 2018). In recent years, research efforts have increasingly focused on the archipelago of Tierra del Fuego, fostering new discussions about the adaptation of HGF populations throughout the Holocene (Lalueza et al. 1997; García-Bour et al. 2004; de la Fuente et al. 2015, 2018; Crespo et al. 2020; Nakatsuka et al. 2020). In this context, genomic studies have provided valuable insights into whether population dynamics were associated with cultural shifts evident in the archaeological record. Uniparental markers analysis showed that populations of Fuego-Patagonia have only C and D

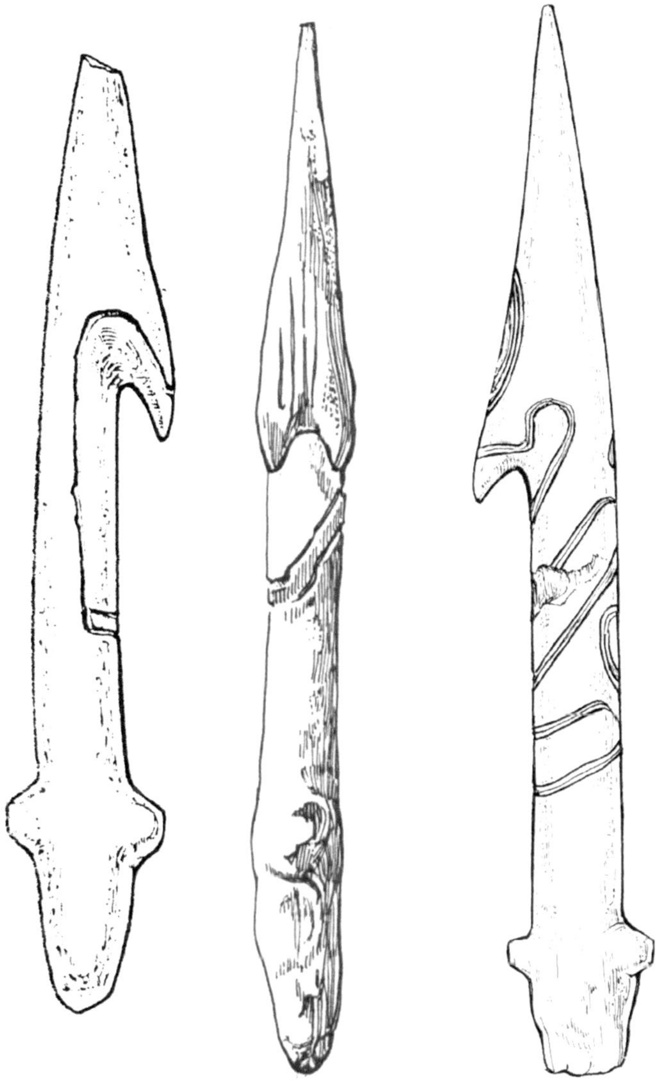

Fig. 2.8 Decorated harpoon points (illustration D. Alonso)

mitochondrial haplogroups and low Y-chromosome diversity. This is consistent with a bottleneck in the founding groups followed by strong genetic drift and isolation, mainly in light of the absence of the founding Native American haplogroups A2 and B2 in ancient populations (Lalueza et al. 1997; García-Bour et al. 2004; de la Fuente 2015). Several genome studies also demonstrated a significant degree of continuity from the archaeological individuals to present-day ones (de la Fuente et al. 2015; Crespo et al. 2020; Nakatsuka et al. 2020).

Regarding spatial and temporal population dynamics along the southern coast of Tierra del Fuego, Crespo et al. (2020) examine maternal genetic diversity in HGF of southern Tierra del Fuego, focusing on the northern coast of the Beagle Channel, Península Mitre, and Isla de los Estados through ancient mitochondrial DNA analysis. Their findings reveal population structuring between the northern coast of the Beagle Channel and the Península Mitre–Isla de los Estados regions, as evidenced by the networks of clades C1 and D1. The C1 haplotypes identified in these areas exhibit affinities with those found in ancient and contemporary Yagan groups. Similarly, individuals carrying the D1g subclade display haplotypes shared with continental populations. Additionally, some individuals from Península Mitre–Isla de los Estados possess haplotypes shared with continental populations, potentially due to a recent common origin with northern Tierra del Fuego or continental Patagonia, or to more recent gene flow events involving populations from northern regions (Crespo et al. 2020). Overall, the genomic data and analyses conducted by Crespo et al. (2020) align with prior archaeological hypotheses suggesting that the southern coast of the Isla Grande de Tierra del Fuego constituted a connected region inhabited by distinct native groups (Zangrando et al. 2009).

Nakatsuka et al. (2020) make a significant contribution to the understanding of South Patagonia by demonstrating how genome studies can identify distinct biological and cultural evolutionary processes associated with material culture changes recorded during the Holocene. The authors examine the covariation between genomic data and technological transitions as cultural markers. Their findings refute the hypothesis of an early large-scale coastal dispersion into the western archipelago of Patagonia. Instead, they suggest that gene flow occurred among marine-adapted populations within the archipelago beginning in the Middle Holocene. The authors identify at least three major north-to-south gene flow events that influenced Fuego-Patagonia. The first occurred at least 6600 cal yr BP, based on the radiocarbon dates of an individual from the Punta Santa Ana site in Península Brunswick. This event may correspond to the stream that brought the ancestors of mid-Holocene individuals, potentially marking an initial colonization event. The authors also observed that both this individual and another from La Arcillosa 2 (5800 cal yr BP), located on the north Atlantic coast of Tierra del Fuego, were genetically equidistant from later groups, including both maritime and terrestrial-reliant populations. Furthermore, both individuals were equally distant from all other American groups outside the region during any time frame analyzed. This genomic similarity suggests that groups with comparable genetic profiles adopted different cultural niches in littoral environments during the Middle Holocene.

The second gene flow introduced ancestry from a lineage associated with the maritime-adapted individual from the Ayayema site (~4700 cal yr BP), replacing lineages related to Punta Santa Ana and La Arcillosa 2. This migration from the northwest might be linked to changes in lithic technology, including the cessation of green obsidian use and the introduction of large bifacial projectile points in the archipelago.

The third gene flow brought ancestry from Central Chile by at least ~ 2000 cal yr BP, possibly connected to various processes during the Late Holocene. While the initial phase in the archipelago was characterized by low diversity in lithic production and foraging activities, subsequent population influxes correspond to diversification in subsistence strategies, including the adoption of new technologies.

Finally, Nakatsuka et al. demonstrate active genetic mixing among spatially proximate Fuegian foragers during the Late Holocene (~2200–1200 cal yr BP). However, this interaction declined over time, resulting in reduced gene flow and cultural differentiation, which may have contributed to the ethnic group configurations observed historically. Similar to the conclusions of Crespo et al. (2020), the authors propose that populations in Península Mitre likely adopted some maritime and terrestrial adaptations through interactions and genetic exchange with neighboring groups.

2.9 Discussion: Linking Material Culture, Landscape and Genes

A critical aspect of understanding cultural niche construction in past human populations lies in analyzing the mechanisms through which information is acquired and transmitted across space and generations (Odling-Smee et al. 2003; Laland and O'Brien 2011). While this process is shaped by socially learned knowledge and cultural inheritance, the transmission and acquisition of knowledge are themselves influenced by pre-existing information derived from genetic evolution, complex ontogenetic processes and/or prior social learning (O'Brien and Laland 2012).

In this chapter, we review the production of technology, the circulation of raw materials and the use of ornaments and decorative motifs on artifacts, as these provide archaeologically detectable evidence of spatial and inter-generational variability in culturally transmitted information (Romano et al. 2021). This review reveals—consistent with archaeological sequences from other regions of the archipelago—that technological innovations and material circulation often appear as discontinuities following prolonged periods of stability, marked by abrupt cultural changes that vary across time and space. In addition, this chapter presents an overview of the genetic data available for HGF from the last 6000 cal yr BP. The evidence from the southern margin of Tierra del Fuego suggests that the cultural niche construction of these populations was shaped by complex interactions between environmental variability and demographic fluctuations throughout the Holocene.

The archaeological sequence of the southern coast of Tierra del Fuego reveals significant fluctuations in population sizes, including periods of notable growth as well as prolonged phases of low population density with marked negative deviations. The first population increase occurred during the Middle Holocene, between 7500 and 6300 cal yr BP. This period is characterized by the formation of sites with extensive archaeological sequences, such as layer D of Túnel I site (Orquera and Piana 1986–87).

The presence of green obsidian in several assemblages from this period suggests connections and the exchange of information between the Beagle Channel area and the region immediately north of the Strait of Magellan, encompassing the Brunswick Peninsula and Otway Sound (Orquera and Piana 1999: 91; Álvarez 2004; Fiore 2006). Such connections between different portions of the Fuegian Archipelago were based on the use of canoes as a pivotal mobility technology, which transformed the water separating islands into "water bridges" (Fiore 2006), which in turn defined a key feature of niche construction.

Decoration on harpoon points is another distinctive feature of the Middle Holocene, interpreted as a mechanism for consolidating interactions between groups. This facilitated the strengthening of reproductive and social ties with regions immediately north of the Strait of Magellan (Fiore 2006). The diversity of decorative motifs observed on harpoon points suggests that their creation took place within a relatively free cultural context, driven by social demands for individual expression (Fiore 2011). Since the majority of harpoon decorations were incised, social learning and imitation likely served as foundational mechanisms for transmitting techno-visual knowledge. However, the variability of motifs also indicates a role for individual expression and innovation (Fiore 2020). In summary, in accordance with the premise that image-making techniques have a lower intrinsic variability than motif repertoires, decorative techniques appear to have changed more slowly than the motifs themselves—supporting the idea that a single technique can yield a broad variety of visual patterns (Fiore 2011).

The population increase identified during the Middle Holocene coincides with the consolidation of an interaction network linking the Beagle Channel to other parts of the archipelago. However, this demographic trend does not necessarily imply cultural homogeneity during this period (e.g., Orquera et al. 2011). The extent to which cultural specialization, loss and diversification stemmed from these social networks remains uncertain (Romano et al. 2021). As Morello et al. (2015) suggest, the circulation of green obsidian may have been part of a broader process in which HGF from different regions monitored changes in resources and populations, sharing information and knowledge about accessing various supplies. Notably, the use of decorative techniques appears to be particularly focused on tools specialized for capturing high-ranking resources, such as marine mammals. Moreover, such high labor investment on harpoon decoration was carried out in spite of the risk of losing or breaking the harpoons while hunting—a process that did happen, along with the similar breakage indexes of non-decorated vs. decorated harpoons (Fiore 2011). According to interpretations of the Tierra del Fuego case by Linares Matás and Lim (2021:21), access to energy-dense marine resources, facilitated by specialized toolsets, may have created a positive feedback loop. This loop supported population growth and enhanced information exchange, which, in turn, fostered greater technological and aesthetic investment. One possible interpretation is that social networks, optimized for information transmission and cultural accumulation, contributed to the resilience of populations facing environmental challenges. These networks would have enabled individuals to rely on diverse contacts and sources of information during the environmentally unstable conditions of the Middle Holocene in Tierra del Fuego. This period was marked by significant changes in forest composition associated with neoglaciations, a major volcanic eruption and geomorphological variations linked to marine transgression.

After 6300 cal yr BP, the frequency of occupations declined, stabilizing over a span of approximately 1500 cal yr BP. Between 4800 and 2000 cal yr BP, the southern coast of Tierra del Fuego experienced its lowest recorded population levels, with evidence suggesting possible population collapses. Despite this demographic downturn, the period shows a persistence in bird bone bead production and decoration in the Beagle Channel and is marked by significant technological changes.

The first instance of technological innovation has been associated with the dispersal of large lanceolate points, which were widely distributed across the Patagonian archipelago. The earliest examples are found in the northernmost sectors of the archipelago, such as level 4 of Puente Quilo 1 (5900–4300 cal yr BP) on Chiloé Island (Ocampo and Rivas 2004; Sierralta Navarro 2024) and strata II to IV of Piedra Azul (5300–4800 cal yr BP) in the Reloncaví Sound (Gaete and Navarro 2004). These points have also been documented on the surface in the Chonos Archipelago (Reyes 2021), and at multiple sites in the Otway Sea, Brunswick Peninsula and Última Esperanza, with chronologies around 4700 cal yr BP (Morello et al. 2002; Legoupil 2003; Legoupil and Pigeot 2009; San Román 2013; Langlais and Huidobro 2023). In the Beagle Channel, their appearance is dated to approximately 4400 cal yr BP (Orquera et al. 1977; Orquera and Piana 1999; Piana et al. 2004).

This technology, while widespread across the archipelago, is temporally restricted and disappears within a relatively short time span. Initially, its presence in layer B at Ponsonby site was interpreted by Laming-Emperaire as evidence of an incursion by pedestrian hunters from the interior. However, Legoupil (2003:115) later reinterpreted this assemblage as a product of terrestrial hunters with a mixed economy.

The possibility of a north-to-south spread of this technology within the archipelago has gained significant attention in recent years. Some researchers propose it reflects the dispersal of maritime hunter-gatherer populations distinct from earlier groups (Morello et al. 2002:164; Ocampo and Rivas 2004), while others argue that the spread of the technological concept occurred without population replacement (Piana and Orquera 2007). In this context, Nakatsuka et al. (2020) identify a second population flow associated with the lineage of the individual recovered at the Ayayema site (~4700 cal yr BP), which appears to have replaced lineages related to earlier individuals recovered from Punta Santa Ana (Brunswick Peninsula) and La Arcillosa 2 (northern Tierra del Fuego). Nakatsuka et al. draw a connection between this population flow, the interruption in the use of green obsidian and the emergence of large lanceolate points in the archipelago. Regardless of the mechanisms behind the spread of this technology, its presence on the southern coast of Tierra del Fuego clearly indicates a connection to other areas of the archipelago, as it has yet to be documented in the interior regions of mainland Patagonia or Tierra del Fuego.

Between 3500 and 3300 cal yr BP, coinciding with the peak intensity of the SWW (Björck et al. 2012), the archaeological record shows a prolonged period of significantly reduced occupation frequency, potentially reflecting a population collapse. Following this downturn, around 2700 cal yr BP, notable changes occurred in both the regional archaeological record and the composition of technological assemblages. One major development was the increasing recurrence of occupations along the southeastern coast of Península Mitre and the first recorded settlements on Isla de los Estados. These changes represent the broadest geographical range of HGF populations employing navigational technology in southern Tierra del Fuego (Horwitz 1993), a period characterized by a substantial decrease in the intensity of SWW (Fig. 2.4). From this period onward, medium-sized lithic points attributed to the Fourth Component of Túnel I are recorded for the first time (Orquera and Piana 1986–87:232). By this time, a considerable morphological diversity in these tools had already emerged. In subsequent assemblages, both the number of specimens—either completed or in the process of being manufactured—and their diversity of forms increased (Orquera and Piana 1999:79; Alvarez 2009). This morphological diversity persisted until the contact period. Legoupil (1989:143–147) and Orquera and Piana (1995:71–73) suggest that these points may have been used as components of daggers or spears, implying that some of the observed morphological variability could be attributed to functional adaptations (Ratto 1994).

According to Piana and Orquera (2007), unlike what happened with the large lanceolate points, there are many detailed parallels in the different shapes of the triangular points—including the frequent asymmetrical fins—with specimens from the archaeological records of other regions of Tierra del Fuego. These authors argue that the appearance of these weapon points in the channel sector may have been the result of contacts with inland areas of the Atlantic coast (Piana and Orquera 2007). There are few records of bifacial knapping in Middle Holocene assemblages in northern Tierra del Fuego (Salemme et al. 2014), while this type of technological production became frequent in the last 2000 cal yr BP due to the presence of weapon points (Morello et al. 2012). Although obsidian reappears frequently in archaeological assemblages north of the Strait of Magellan during this period -and sporadically south of the strait (Morello et al. 2015)—there is no evidence that its distribution extended to the southern coast of Tierra del Fuego during the Late Holocene. On the other hand, silicified Miraflores tuff has been recorded in archaeological assemblages from Península Mitre, although always in low frequency. Given its absence in the central Beagle Channel (Borrazzo et al. 2019), its presence in southeastern Tierra del Fuego points to possible vectors of human circulation and social interaction linking this area with northern Tierra del Fuego (Pallo and Borrazzo 2016).

After 2700 cal yr BP there are morphological changes in the detachable harpoon points and they cease to be decorated (Fiore 2011). Although this implies stylistic variations and variations in the technological production process, it does not necessarily imply changes in their functioning or effectiveness (Orquera et al. 2011). During the late Holocene, however, the decoration of bone artifacts is maintained in beads and a few other artifacts. In contrast to the case of harpoon points, designs on beads persist throughout the Holocene. The motifs are simple, fairly standardized and have a low repertoire variability. In terms of their production, this implies that they were subject to greater social control and/or greater force of custom. The identity expressed through these motifs may have extended beyond the Beagle Channel populations to include groups from northern Tierra del Fuego and the Strait of Magellan

(Ortiz-Troncoso 1979; Massone 1988; Massone et al. 1993; Legoupil 2003; Fiore 2011). This is evidenced by the presence of beads with designs highly similar to those recorded in the Beagle Channel (Fiore 2006).

After 2000 cal yr BP, the marked increase in the frequency of radiocarbon dates reflects a significant rise in population size of HGF along the southern coast of Tierra del Fuego. Two distinct periods of positive deviation are notable: one around 1800 cal yr BP and another approximately 500 cal yr BP. The latter represents the most substantial population increase in the entire sequence, as evidenced by the archaeological record from both the Beagle Channel and the southeastern coast of the Península Mitre. However, during this period of population growth, beginning around 1200 cal yr BP, evidence of occupations on Isla de los Estados ceased. This hypothesized abandonment has been interpreted as a possible displacement of marine HGF, potentially linked to increased occupation by other groups on Península Mitre (Horwitz 1990, 1993). This process remains under investigation (Vázquez 2024; Zangrando et al. 2024). This population growth also coincides with two notable climate anomaly events during the last millennium: the Medieval Warm Period or Medieval Climate Anomaly (1100–800 cal yr BP) and the Little Ice Age (600–100 cal yr BP) (Moreno et al. 2014). Genetic studies suggest that the late Holocene was characterized by the greatest biological exchange and interaction between neighboring populations in Tierra del Fuego (Crespo et al. 2020; Nakatsuka et al. 2020). Crespo et al. (2020) document significant genetic exchange between the Beagle Channel and Península Mitre during the late Holocene, while Nakatsuka et al. (2020) identified the highest levels of interaction between 2200 and 1200 cal yr BP. After this period, gene flow diminished, leading to greater cultural differentiation in historical times. In conjunction with this, it is noteworthy that over the last 500 years, the Beagle Channel has experienced a more pronounced population decline compared to the Península Mitre.

2.10 Conclusion

The construction of the cultural niche of the HGF populations inhabiting the southern coast of Tierra del Fuego evolved through various dynamics throughout the Holocene. Based on the review presented in this chapter, which integrates technological and genetic evidence, it appears that during the middle Holocene, there were stronger connections and interactions with populations inhabiting more northerly regions of the archipelago. Later in the sequence, after approximately 2700 cal yr BP, the technological and genetic record suggests that the construction of the cultural niche was shaped by more localized and fluid social interactions, largely confined to the area of Tierra del Fuego.

References

Acevedo A (2024) Arte rupestre y construcción de nicho: una teoría, un caso y una hipótesis. Estud Atacameños 70:e6121. https://doi.org/10.22199/issn.0718-1043-2024-0022

Alberti J, Fernández MV (2015) Propuesta clasificatoria para las materias primas líticas en Patagonia (Argentina). Arqueología 21(2):211–235. https://doi.org/10.34096/arqueologia.t21.n2.223

Álvarez M, Briz i Godino I, Pal N, Salvatelli L (2010) Contextos de uso y diseños: una propuesta metodológica para el análisis de la variabilidad de los conjuntos líticos. In: Bárcena R, Chiavazza H (eds) Actas del XII Congreso Nacional de Arqueología Argentina, Tomo I/V. Facultad de Filosofía y Letras UNCUYO. Instituto de Ciencias Humanas, Sociales y Ambientales, CONICET, Mendoza p 67–72

Álvarez M, Zurro D, Briz i Godino I, Madella M, Osterrieth M, Borrelli N (2009) Análisis de los procesos productivos en las sociedades cazadoras-recolectoras-pescadoras de la costa norte del canal Beagle (Argentina): el sitio Lanashuaia. In: Salemme M, Santiago F, Álvarez M, Piana EL, Vázquez MM, Mansur M (eds) Arqueología de Patagonia: Una mirada desde el último confín. Utopías, Ushuaia, pp 903–917

Álvarez M (2004) Estrategias tecnológicas en los grupos canoeros tempranos del área fuegopatagónica. Magallania 32:191–208. https://www.magallania.cl/index.php/magallania/article/view/1208

Álvarez M (2009) Tendencias y cambios en las prácticas tecnológicas de los grupos cazadores recolectores del extremo sur sudamericano. In: Salemme M, Santiago F, Álvarez M, Piana EL, Vázquez MM, Mansur M (eds) Arqueología de Patagonia: Una mirada desde el último confín. Utopías, Ushuaia, pp 19–33

Álvarez M (2011) Puntas de arma del extremo sur de patagonia: Algunas consideraciones sobre diseño y contexto de uso. In: Martínez JG, Bozzutto DL (eds) Armas prehispánicas: Múltiples enfoques para su estudio en Sudamérica. Fundación de Historia Natural Félix de Azara, Buenos Aires pp 15–36

Andrefsky W (1994) Raw-material availability and the organization of technology. Am Antiq 59(1):21–34. https://doi.org/10.2307/3085499

De Angelis H (2015) Arqueología de los cazadores-recolectores de la faja central de la isla Grande de Tierra del Fuego. Sociedad Argentina de Antropología, Buenos Aires

Bailey GN (1983) Problems of site formation and the interpretation of spatial and temporal discontinuities in the distribution of coastal middens. In: Masters PM, Flemming NC (eds) Quaternary coastlines and marine archaeology. Academic Press, London, pp 559–582

Bailey GN (2014) New developments in submerged prehistoric archaeology: an overview. In: Evans AM, Flatman JC, Flemming NC (eds) Prehistoric archaeology on the continental shelf. A Global Review Springer Nature, New York, pp 291–300

Barceló JA, Piana EL, Martinioni DR (2002) Archaeological spatial modelling: a case study from Beagle Channel (Argentina). In: Burenhult G, Arvidsson J (eds) Archaeological Informatics: pushing the envelope, computer applications and quantitative methods in archaeology. BAR Series 1016, Oxford, pp 351–360

Binford LR (1982) The archaeology of place. J Anthropol Archaeol 1:5–31. https://doi.org/10.1016/0278-4165(82)90006-X

Bird J (1993) Viajes y arqueología en Chile Austral. Ediciones de la Universidad de Magallanes, Punta Arenas

Bjerck HB, Zangrando AF, Breivik HM, Piana EL, Negre J (2016) Marine ventures: the cambaceres surveys (2009, 2011–2013). NTNU Vitenskapsmuseetarkeologisk rapport. http://www.ntnu.no/vitenskapsmuseet/publikasjoner

Björck S, Rundgren M, Ljung K, Unkel I, Wallin A (2012) Multy-proxy analysis of a peat bog on Isla de los Estados, easternmost Tierra del Fuego: a unique record of the variable Southern Hemisphere Westerlies since the last deglaciation. Quat Sci Rev 42:1–14. https://doi.org/10.1016/j.quascirev.2012.03.015

Björck S, Lambeck K, Moller P, Waldmann N, Bennike O, Jiang H, Li D, Sandgren P, Nielsen AB, Porter CT (2021) Relative Sea level changes and glacio-isostatic modelling in the beagle channel, Tierra Del Fuego, Chile: Glacial and Tectonic Implications. Quat Sci Rev 251(1):106657. https://doi.org/10.1016/j.quascirev.2020.106657

Borrazzo K, D'Orazio M, Etchichury MC (2010) Distribución espacial y uso prehistórico de las materias primas líticas del chorrillo Miraflores en el norte de Isla Grande de Tierra del Fuego (Argentina). Rev Chil Antropol 22:77–97

Borrazzo K, Morello F, Borrero LA, D'Orazio M, Etchichury MC, Massone M, De Ángelis H (2015) Caracterización y uso de las materias primas líticas de Chorrillo Miraflores en el extremo meridional de Fuego-Patagonia. Intersecc Antropol 16(1):155–167

Borrazzo KB, Borrero LA, Pallo MC (2019) Exploring lithic transport in Tierra del Fuego (Southern South America). J Archaeol Sci Rep 24:220–230. https://doi.org/10.1016/j.jasrep.2019.01.002

Borrero LA (1994–95) Arqueología de la Patagonia. Palimpsesto 4:9–69

Borrero LA (2014) Multi-service taphonomy. Shells, garbage, and floating palimpsests. Intersecc Antropol 15:13–20

Borrero LA, Martin F, Barberena R (2011) Visits, "Fuegians", and information networks. In: Whallon R, Lovis W, Hitchcock R (eds) Information and its role in hunter-gatherer bands. University of California Press, Los Angeles, pp 249–269

Borromei AM, Coronato A, Franzén LG, Ponce JF, López Sáez JA, Maidana N, Rabassa J, Candel MA (2010) Multiproxy record of Holocene paleoenvironmental change, Tierra del Fuego, Argentina. Palaeogeogr Palaeoclim Palaeoecol 286:1–16. https://doi.org/10.1016/j.palaeo.2009.11.033

Borromei AM, Ponce JF, Coronato A, Candel S, Olivera D, Masaaki O (2014) Postglacial vegetation reconstruction and its relation with the relative sea-level rise in the E extreme of Beagle Channel, Tierra del Fuego. Argentina. Andean Geol 41(2):362–379. https://doi.org/10.5027/andgeoV41n2-a05

Borromei AM, Quattrocchio M (2007) Holocene sea-level change inferred from palynological data in the Beagle Channel, southern Tierra del Fuego, Argentina. Ameghiniana 44(1):161–171. https://www.ameghiniana.org.ar/index.php/ameghiniana/article/view/681

Borromei AM, Musotto LL, Coronato A, Ponce JF, Pontevedra Pombal X (2016) Postglacial vegetation and climate changes inferred from a peat pollen record in the Río Pipo valley, southern Tierra del Fuego. In: Martínez M, Olivera D (eds) Palinología del Meso-Cenozoico de Argentina - Volumen en homenaje a Mirta Elena Quattrocchio. Publ Electrón Asoc Paleontol Argent 16(2):168–183. https://doi.org/10.5710/PEAPA.24.03.2016.91

Bridges EL (2000 [1947]) The uttermost part of the earth. Hodder and Stoughton, London

Broecker WS (2001) Was the medieval warm period global? Sci 291:1497–1499. https://doi.org/10.1126/science.291.5508.1497

Bujalesky G (1998) Holocene coastal evolution of Tierra del Fuego, Argentina. Quat S Am Antarct Penins 11:247–281

Bujalesky G (2007) Coastal geomorphology and evolution of Tierra del Fuego (Southern Argentina). Geol Acta 5(4):337–362

Bujalesky G (2011) The flood of the Beagle Valley (11,000 yr B.P.) Tierra del Fuego. Ana Inst Patag 39(1):5–21. https://www.analesdelinstitutodelapatagonia.cl/index.php/analespatagonia/article/view/225

Candel MS, Borromei AM, Louwye S (2018) Early to middle holocene palaeoenvironmental reconstruction of the beagle channel (Southernmost Argentina) based on terrestrial and marine palynomorphs. Boreas 47(4):1072–1083. https://doi.org/10.1111/bor.12322

Chapman A (1986) Los Selk'nam. La vida de los Onas. Emecé, Buenos Aires

Chapman A (1987) La Isla de los Estados en la prehistoria. Primeros datos arqueológicos. Eudeba, Buenos Aires

Charlin J, Cardillo M, Borrazzo K (2014) Spatial patterns in Late Holocene lithic projectile point technology of Tierra del Fuego (Southern South America): assessing size and shape changes. World Archaeol 46(1):78–100. https://doi.org/10.1080/00438243.2014.890914

Charlin J, Augustat C, Urban C (2016) Metrical variability in ethnographic arrows from southernmost Patagonia: comparing collections from Tierra del Fuego at European museums. J Anthropol Archaeol 41:313–326. https://doi.org/10.1016/j.jaa.2016.01.006

Conkey M (1980) The identification of prehistoric hunter-gatherer aggregation sites: the case of Altamira. Curr Anthropol 21(5):609–630. https://doi.org/10.1086/202540

Coronato A, Borromei A, Ponce JF, Candel S, Musotto L, Fernández M, Laprida C, Mehl A, Montes A, San Martin C, Savoretti A, Cusminsky G, Gordillo S, Orgeira MJ (2022) Holocene environmental changes in the fuegian forest and steppe. Argentina. J South Am Earth Sci 119:103952. https://doi.org/10.1016/j.jsames.2022.103952

Coronato A, Coronato F, Mazzoni E, Vázquez M (2008) The physical geography of Patagonia and Tierra del Fuego. In: Rabassa J (ed) Late Cenozoic of Patagonia and Tierra del Fuego. Dev Quat Sci 11:13–55. https://doi.org/10.1016/S1571-0866(07)10003-8

Crema ER, Bevan A (2021) Inference from large sets of radiocarbon dates: software and methods. Radiocarbon 63(1):23–39. https://doi.org/10.1017/RDC.2020.95

Crema ER, Bevan A, Shennan S (2017) Spatio-temporal approaches to archaeological radiocarbon dates. J Archaeol Sci 87:1–9. https://doi.org/10.1016/j.jas.2017.09.007

Crespo CM, Cardozo DG, Tessone A, Vázquez MM, Kisielinski C, Arencibia V, Tackney J, Zangrando AF, Dejean CB (2020) Distribution of maternal lineages in hunter-gatherer societies of the southern coast of Tierra del Fuego, Argentina. Am J Phys Anthropol 173(4):709–720. https://doi.org/10.1002/ajpa.24107

De Angelis H (2014) Aproximación experimental en la utilización de tobas silicificadas en la costa sur del lago Fagnano Tierra Del Fuego. Comechingonia 8(1):1–19

de la Fuente C, Galimany J, Kemp BM, Judd K, Reyes O, Moraga M (2015) Ancient marine hunter-gatherers from Patagonia and Tierra del Fuego: Diversity and differentiation using uniparentally inherited genetic markers. Am J Phys Anthropol 158(4):719–729. https://doi.org/10.1002/ajpa.22815

de la Fuente C, Ávila-Arcos MC, Galimany J, Carpenter ML, Homburger JR, Blanco A, Contreras P, Cruz Dávalos D, Reyes O, San Roman M, Moreno-Estrada A, Campos PF, Eng C, Huntsman S, Burchard EG, Malaspinas AS, Bustamante CD, Willerslev E, Llop E, Verdugo RA, Moraga M (2018) Genomic insights into the origin and diversification of late maritime hunter-gatherers from the Chilean Patagonia. Proc Natl Acad Sci USA 115(17):4006–4012. https://doi.org/10.1073/pnas.1715688115

Eidshaug JSP, Bjerck HB, Breivik HM, Risbøl O, Tivoli AM, Zangrando AF (2024) From earth to sky: Large-scale archaeological settlement patterns in southernmost South America based on ground surveys, UAV LiDAR, and open access satellite imagery. J Isl Coast Archaeol, pp 1–40. https://doi.org/10.1080/15564894.2024.2426828

Emperaire J, Laming A (1961) Les Gisements Archeólogiques des Ilas Englefield et Vivian dans la Mer d'Otway. J Soc Américanistes 50:7–75. https://www.jstor.org/stable/24604048

Esper J, Cook ER, Schweingruber FH (2002) Low-frequency signals in long tree-ring chronologies for reconstructing past temperature variability. Sci 295:2250–2253. https://doi.org/10.1126/science.1066208

Estévez J, Vila A (1995) Encuentros en los conchales fueguinos. Treballs D'Etnoarqueologia 1. CSIC, Universidad Autónoma de Barcelona, Barcelona

Estévez J, Vila A (2006) Variability in the lithic and faunal record through 10 reoccupations of a XIX century Yamana Hut. J Anthropol Archaeol 25(4):408–423. https://doi.org/10.1016/j.jaa.2006.03.007

Estévez J, Vila A (2007) Twenty years of ethnoarchaeological research in Tierra del Fuego: some thoughts for European shell-midden archaeology. In: Craig OE, Bailey GN (eds) Milner N. Oxbow Books, Oxford, pp 183–195

Fernández M, Ponce FJ, Zangrando AF, Borromei AM, Musotto LL, Alunni DV, Vázquez M (2020) Relationships between terrestrial animal exploitation, marine hunter-gatherers and palaeoenvironmental conditions during the Middle-Late Holocene in the Beagle Channel region (Tierra del Fuego). Quat Int 549:208–217. https://doi.org/10.1016/j.quaint.2018.05.032

Fiore D (2011) Art in time. Diachronic rates of change in the decoration of bone artefacts from the Beagle Channel region (Tierra del Fuego, Southern South America). J Anthropol Archaeol 30:484–501. https://doi.org/10.1016/j.jaa.2011.07.002
Fiore D (2006) Puentes de agua para el arte mobiliar: la distribución espacio temporal de artefactos óseos decorados en Patagonia meridional y Tierra del Fuego. Cazadores-Recolectores del Cono Sur. Rev Arqueol I, pp 137–147
Fiore D (2014) Pinturas corporales fueguinas. Una arqueología visual. In: Oria J, Tivoli A (eds) Cazadores de mar y tierra. Estudios recientes en arqueología fueguina. Editorial Cultural Tierra del Fuego, Ushuaia, pp 409–433
Fiore, D (2020) The art of making images. Technological affordance, design variability and labour organization in the production of engraved artefacts and body paintings in Tierra del Fuego (Southern South America). J Archaeol Method Theory 27(3):481–510. https://doi.org/10.1007/s10816-020-09474-7
Fleming K, Johnston P, Zwartz D, Yokoyama Y, Lambeck K, Chappell J (1998) Refining the Eustatic sea-level Curve since the Last Glacial Maximum Using Far- and intermediate-field Sites. Earth Planet Sci Lett 163(1–4):327–342. https://doi.org/10.1016/S0012-821X(98)00198-8
Franch Bach A (2024) Management of plant resources in southern hunter-gatherer societies: a review of archaeobotanical evidence of Isla Grande of Tierra del Fuego (Argentina). Veget Hist Archaeobot. https://doi.org/10.1007/s00334-024-01031-6
Franch Bach A, Capparelli A, Piqué R, Zangrando AF (2025) Early-Mid Holocene archaeobotanical record of the southern coast of Tierra del Fuego. Environ Archaeol. https://doi.org/10.1080/14614103.2025.2532992
Gaete N, Navarro X (2004) Estrategias de vida de canoeros cazadores pescadores recolectores del seno de Reloncaví: entre el bosque siempreverde y el Mar Interior. Región de Los Lagos, Chile. In: Civalero MT, Fernández PM, Guráieb AG (eds) Contra Viento y Marea. Arqueología de Patagonia. INAPL - Sociedad Argentina de Antropología, Buenos Aires, pp 217–234
Gallardo F, Cabello G, Sepúlveda M, Ballester B, Fiore D, Prieto A (2023) Yendegaia Rockshelter, the first rock art site on Tierra del Fuego Island and social interaction in Southern Patagonia (South America). Lat Am Antiq 34(3):532–549. https://doi.org/10.1017/laq.2022.47
Gamble C (1982) Interaction and alliance in Palaeolithic society. Man 17:92–107. https://doi.org/10.2307/2802103
García-Bour J, Pérez-Pérez A, Alvarez S, Fernández E, López-Parra A, Arroyo-Pardo E, Turbón D (2004) Early population differentiation in extinc aborigines from Tierra del Fuego-Patagonia: ancient mtDNA sequence and Y-chromosome STR characterization. Am J Phys Anthropol 123(4):361–370. https://doi.org/10.1002/ajpa.10337
Gordillo S, Bujalesky G, Pirazzoli PA, Rabassa J, Saliège JF (1992) Holocene raised beaches along the northern coast of the Beagle Channel, Tierra del Fuego, Argentina. Palaeogeogr Palaeoclim Palaeoecol 99:41–54. https://doi.org/10.1016/0031-0182(92)90006-Q
Gordillo S, Coronato AMJ, Rabassa JO (1993) Late Quaternary evolution of a subantarctic paleofjord, Tierra del Fuego. Quat Sci Rev 12:889–897. https://doi.org/10.1016/0277-3791(93)90027-J
Gusinde M (1990 [1937]) Los indios de Tierra del Fuego. Volume I (I). Los selk'nam. Centro Argentino de Etnología Americana, Buenos Aires
Haas R, Kuhn SL (2019) Forager mobility in constructed environments. Curr Anthropol 60 (4): 499-535
Hernandez Herrero O (2024) Variabilidad de las prácticas tecnológicas líticas en Heshkaia 60 durante el Holoceno medio (costa sureste del canal Beagle, Tierra del Fuego). Paper presented at the XII Jornadas de Arqueología de la Patagonia 2024, Universidad Nacional de la Patagonia Austral, Río Gallegos, 21–25 October 2024
Heusser CJ (1989) Climate and chronology of Antarctica and adjacent South America over the past 30,000 yr. Palaeogeogr Palaeoclim Palaeoecol 76:31–37. https://doi.org/10.1016/0031-0182(89)90101-6

Heusser CJ (1998) Deglacial paleoclimate of the American sector of the Southern Ocean: Late Glacial-Holocene records from the latitude of Canal Beagle (55ºS). Argentine Tierra del Fuego. Palaeogeogr Palaeoclim Palaeoecol 141:277–301. https://doi.org/10.1016/S0031-0182(98)00053-4

Heusser CJ (2003) Ice Age Southern Andes–A Chronicle of Paleoecological Events. Developments in Quaternary Science 3. Elsevier, Amsterdam

Hogg AG, Heaton TJ, Hua Q, Palmer JG, Turney CSM, Southon J, Bayliss A, Blackwell PG, Boswijk G, Bronk Ramsey G, Pearson C, Petchey F, Reimer P, Reimer R, Wacker L (2020) SHCal20 Southern Hemisphere Calibration, 0–55,000 Years cal BP. Radiocarbon 62(4):759–778. https://doi.org/10.1017/RDC.2020.59

Horwitz V (1993) Maritime settlement patterns: the case from Isla de los Estados (Staten Island). Arqueol Contemp 4:163–176

Horwitz V, Weissel M (2011) Arqueología de Isla de los Estados: La frontera de la abundancia. In: Zangrando AF, Vázquez M, Tessone A (eds) Los cazadores recolectores del extremo oriental fueguino. Arqueología de Península Mitre e Isla de los Estados. Sociedad Argentina de Antropología, Buenos Aires, pp 143–170

Horwitz V (1990) Maritime settlement patterns in Southeastern Tierra del Fuego (Argentina). PhD Dissertation, University of Kentucky

Hyades P, Deniker J (2007 [1891]) Los Indígenas según Hyades y Deniker. In: Legoupil D, Prieto, A (eds.) Etnografía de los indios Yaghan en la Misión Científica. Ediciones Universidad de Magallanes—Instituto Francés de Estudios Andinos, Punta Arenas, pp 69–329

Jochim M (1983) Palaeolithic cave art in ecological perspective. In: Bailey G (ed) Hunter-gatherer economy in prehistory. Cambridge University Press, Cambridge, New Directions in Archaeology, pp 212–219

Laland KN, O'Brien MJ (2011) Cultural niche construction: an introduction. Biol Theory 6:191–202. https://doi.org/10.1007/s13752-012-0026-6

Lalueza C, Pérez-Pérez A, Prats E, Cornudella L, Turbón D (1997) Lack of founding amerindian mitochondrial DNA lineages in extinct aborigines from Tierra Del Fuego-Patagonia. Hum Mol Genet 6(1):41–46. https://doi.org/10.1093/hmg/6.1.41

Lambeck K, Rouby H, Purcell A, Sun Y, Sambridge M (2014) Sea level and global ice volumes from the Last Glacial Maximum to the Holocene. Proc Natl Acad Sci USA 111(43). https://doi.org/10.1073/pnas.1411762111

Langlais M, Huidobro C (2023) El equipamiento de piedra tallada del sitio Offing 2 (Locus 1). In: Legoupil D (ed) Entre Patagonia y Tierra del Fuego: Los nómadas del mar del islote Offing (I. Dawson–Estrecho de Magallanes). Ediciones Universidad de Magallanes, Colección Poblamiento Humano de Fuego-Patagonia, Punta Arenas, pp 69–102

Legoupil D, Pigeot N (2009) Les grandes pointes foliacées du type Ponsonby: un traceur culturel en Patagonie australe. J Soc Amér 95(2):7–32. https://doi.org/10.4000/jsa.10941

Legoupil D, Fontugne M (1997) El poblamiento marítimo de los archipiélagos de Patagonia: núcleos antiguos y dispersión reciente. Anal Inst Patag Ser Cs Hum 25:75–87. https://hal.science/hal-04819150v1

Legoupil D (1989) Ethno-archéologie dans les Archipels de Patagonie: les Nomades Marins de Punta Baja. Éd. Recherches sur les Civilisations, Ministère des Affaires etrangères Mémoire de l'ADPF

Legoupil D (2003) Cazadores-recolectores de Ponsonby (Patagonia austral) y su paleoambiente desde VI al III milenio AC. Magallania. Tirada Especial Documentos 31, Punta Arenas

Legoupil D (2023) Entre Patagonia y Tierra del Fuego. Los nómadas del mar del Islote Offing (I. Dawson - Estrecho de Magallanes). Colección Poblamiento Humano del Fuego-Patagonia, Ediciones Universidad de Magallanes, Puntas Arenas

Linares Matás GJ, Lim SJ (2021) "This is the way": Knowledge networks and toolkit specialization in the circumpolar coastal landscapes of western Alaska and Tierra del Fuego. J Isl Coast Archaeol 19(1):1–29. https://doi.org/10.1080/15564894.2021.2000073

Llamas B, Fehren-Schmitz L, Valverde G, Soubrier J, Smith C, Reich D, Ho SYW, Cooper A, Haak W (2016) Ancient mitochondrial DNA provides high-resolution time scale of the peopling of the Americas. Sci Adv 2:e1501385. https://doi.org/10.1126/sciadv.1501385
Mansilla CA, McCulloch RD, Morello F (2016) Palaeoenvironmental change in Southern Patagonia during the Lateglacial and Holocene: Implications for forest refugia and climate reconstructions. Palaeogeogr Palaeoclim Palaeoecol 447:1–11. https://doi.org/10.1016/j.palaeo.2016.01.041
Mansur ME, Piqué R (2009) Between the forest and the sea: hunter-gatherer occupations in the subantarctic forests in Tierra del Fuego Argentina. Arctic Anthropol 46(1–2):144–157. https://doi.org/10.1353/arc.0.0022
Markgraf V (1983) Late and postglacial vegetational and paleoclimatic changes in subantarctic, temperate and arid environments in Argentina. Palynology 7:43–70. https://doi.org/10.1080/01916122.1983.9989252
Massone M (1988) Artefactos óseos del yacimiento arqueológico Tres Arroyos (Tierra del Fuego). Anal Inst Patag 8:107–112
Massone M, Jackson D, Prieto A (1993) Perspectiva arqueológica de los Selk´nam. Centro de Investigaciones Diego Barros Arana, Santiago de Chile
Mauquoy D, Blaauw B, van Geel B, Borromei A, Quattrocchio M, Chambers FM, Possnert G (2004) Late Holocene climatic changes in Tierra del Fuego based on multiproxy analysis of peat deposits. Quat Res 61:148–158. https://doi.org/10.1016/j.yqres.2003.10.001
McCulloch RD, Mansilla CA, Morello F, De Pol-Holz R, San Román M, Tisdall E, Torres J (2019) Late glacial and holocene landscape change and rapid climate and coastal impacts in the canal beagle Southernmost Patagonia. J Quat Sci 34(8):674–684. https://doi.org/10.1002/jqs.3167
McCulloch RD, Blaikie J, Jacob B, Mansilla CA, Morello F, De Pol-Holz R, San Román M, Tisdall E, Torres J (2020) Late Glacial and Holocene climate variability, southernmost Patagonia. Quat Sci Rev 229:106131. https://doi.org/10.1016/j.quascirev.2019.106131
Meltzer D (1989) Was stone exchange among eastern North American Paleoindians? In: Ellis C, Lothrop J (eds) Eastern Paleoindian Lithic Resource Use. Westview Press, Boulder, pp 11–39
Menounos B, Clague JJ, Osborn G, Thompson Davies P, Ponce JF, Goehring B, Maurer M, Rabassa J, Coronato A, Marr R (2013) Latest Pleistocene and Holocene glacier fluctuations in southernmost Tierra del Fuego, Argentina. Quat Sci Rev 77:70–79. https://doi.org/10.1016/j.quascirev.2013.07.008
Mercer JH (1968) Variations of some Patagonian glaciers since the Late-Glacial. Am J Sci 266:91–109
Mercer JH (1982) Holocene glacier variations in southern South America. Striae 18:35–40
Moore DM (1983) Flora of Tierra del Fuego. Nelson, Oswestry
Morello F, San Román M, Prieto A (2002) Puntas de proyectil lanceoladas en Patagonia Meridional y Tierra del Fuego. Anal Ins Patag Ser Cs Hum 30:155–166
Morello F, Borrero LA, Massone M, Stern C, Garcia-Herbst A, McCulloch R, Arroyo-Kalin M, Calás E, Torres J, Prieto A, Martinez I, Bahamonde G, Cárdenas P (2012) Hunter-gatherers, biogeographic barriers and the development of human settlement in Tierra del Fuego. Antiq 86:71–87. https://doi.org/10.1017/S0003598X00062463
Morello F, Stern CH, San Román M (2015) Obsidiana verde en Tierra del Fuego y Patagonia: caracterización, distribución y problemáticas culturales a lo largo del Holoceno. Intersecc Antropol, Vol Esp 2:139–153
Moreno PI, Vilanova I, Villa-Martínez R, Garreaud RD, Rojas M, De Pol-Holz R (2014) Southern annular mode-like changes in southwestern Patagonia at centennial timescales over the last three millennia. Nat Commun 5:4375. https://doi.org/10.1038/ncomms5375
Moreno PI, Vilanova I, Villa-Martínez R, Dunbar RB, Mucciarone DA, Kaplan MR, Garreaud RD, Rojas M, Moy CM, De Pol-Holz R (2018) Onset and evolution of southern annular mode-like changes at centennial timescale. Sci Rep 8:3458. https://doi.org/10.1038/s41598-018-21836-6
Mörner NA (1987) Sea level changes and tectonics in tierra del fuego. Bull Int Union Quat Res Neotectonics Comm 10:31

Mörner NA (1991) Holocene sea level changes in the Tierra del Fuego region. Bol Inst Geociênc Univ São Paulo Spec Publ 8:133–151. https://doi.org/10.11606/issn.2317-8078.v0i8p133-151
Musotto LL, Borromei AM, Coronato AMJ, Menounos B, Osborn G, Marr R (2016) Late Pleistocene and Holocene palaeoenvironmental changes in central Tierra del Fuego (~54°S) inferred from pollen analysis. Veg Hist Archaeobot 25:117–130. https://doi.org/10.1007/s00334-015-0537-8
Musotto LL, Borromei AM, Bianchinotti MV, Coronato AMJ, Menounos B, Osborn G, Marr R (2017) Postglacial environments in the southern coast of Lago Fagnano, central Tierra del Fuego, Argentina, based on pollen and fungal microfossils analyses. Rev Palaeobot Palynol 238:43–54. https://doi.org/10.1016/j.revpalbo.2016.11.016
Nakatsuka N, Luisi P, Motti JMB, Salemme M, Santiago F, D'Angelo del Campo MD, Vecchi RJ, Espinosa-Parrilla Y, Prieto A, Adamski N, Lawson AM, Harper TK, Culleton BJ, Kennett DJ, Lalueza-Fox C, Mallick S, Rohland N, Guichón RA, Cabana GS, Nores R, Reich D (2020) Ancient genomes in South Patagonia reveal population movements associated with technological shifts and geography. Nat Commun 11:3868. https://doi.org/10.1038/s41467-020-17656-w
Nelson M (1991) The study of technological organization. J Archaeol Method Theory 3:57–100. https://www.jstor.org/stable/20170213
Nowell A (2017) Visual cultures in the Upper Palaeolithic. Camb Archaeol J 27(4):599–606. https://doi.org/10.1017/S0959774317000634
Obelic B, Álvarez A, Argullós J, Piana EL (1998) Determination of water paleotemperature in the Beagle Channel (Argentina) during the last 6000 yr through stable isotope composition of Mytilus edulis shells. Quat S Am Antarct Penins 11:47–71
O'Brien MJ and Laland KN (2012) Genes, culture, and agriculture. An example of human niche construction. Curr Anthropol (53) 4: 434-470
Ocampo C, Rivas P (2004) Poblamiento temprano de los extremos geográficos de los canales patagónicos: Chiloé e Isla Navarino 1. Chungara 36(E1):317–331. https://doi.org/10.4067/S0717-73562004000300034
Odling-Smee FJ, Laland KN, Feldman MW (2003) Niche construction: the neglected process in evolution. Princeton University Press, Princeton
Orquera LA, Piana EL (1987) Human littoral adaptation in the Beagle Channel region: the maximum possible age. Quat S Am Antarct Penins 5:133–165
Orquera LA, Piana EL (2006) El poblamiento inicial del Área Litoral Sudamericana Sudoccidental. Magallania 34:21–37. https://doi.org/10.4067/S0718-22442006000200003
Orquera LA, Piana EL (2009) Sea nomads of the Beagle Channel in Southernmost South America: over six thousand years of coastal adaptation and stability. J Isl Coast Archaeol 4:61–81. https://doi.org/10.1080/15564890902789882
Orquera LA, Legoupil D, Piana EL (2011) Littoral adaptation at the southern end of South America. Quat Int 239(1–2):61–69. https://doi.org/10.1016/j.quaint.2011.02.032
Orquera LA, Piana EL (1995) Túnel VII: la excavación. In: Estévez J, Vila A (eds) Encuentros en los conchales fueguinos. Treballs d'Etnoarqueología n°1, CSIC—Universidad Autónoma de Barcelona, Madrid-Barcelona, pp 47–81
Orquera LA, Piana EL (1996) El sitio Shamakush I (Tierra del Fuego, República Argentina). Rel Soc Arg Antropol XXI:215–265
Orquera LA, Piana EL (1999) Arqueología de la región del canal Beagle (Tierra del Fuego, República Argentina). Sociedad Argentina de Antropología, Buenos Aires
Orquera LA, Piana EL (2005) La adaptación al litoral sudamericano sudoccidental: qué es y quiénes, cuándo y dónde se adaptaron. Rel Soc Arg Antropol XXX:11–32
Orquera LA, Piana EL (2020) Calibración radiocarbónica en la región del canal Beagle. Rel Soc Arg Antropol XLV(1):201–221
Orquera LA, Piana EL (1989–90) La formación de los montículos arqueológicos de la región del canal Beagle. Runa XIX:59–82. https://doi.org/10.34096/runa.v19i1.4271
Orquera LA, Piana EL (1986–87) Composición topológica y datos tecnomorfológicos y tecnofuncionales de los distintos conjuntos arqueológicos del sitio Túnel I (Tierra del Fuego). Rel Soc Arg Antropol XVII:201–239

Orquera LA, Sala A, Piana EL, Tapia A (1977) Lancha Packewaia. Arqueología de los Canales Fueguinos. Temas de Arqueología. Huemul, Buenos Aires

Orquera LA (2022). El poblamiento temprano de Tierra del Fuego. Intersecc Antropol 23(1):141–151. https://doi.org/10.37176/iea.23.1.2022.654

Ortiz-Troncoso O (1973) Aspectos Arqueológicos de la Península de Brunswick. Anal Inst Patag 6:109–130

Ortiz-Troncoso O (1975) Los Yacimientos de Punta Santa Ana y Bahía Buena (Patagonia Austral). Excavaciones y Fechados Radiocarbónicos. Anal Inst Patag 7:93–122

Ortiz-Troncoso O (1979) Punta Santa Ana et Bahía Buena: deux Gisements sur une ancienne ligne de rivage dans le détroit de Magellan. J Soc Amér 60(6):133–204. https://www.jstor.org/stable/24604458

Pallo MC (2016) Obsidian transport in Fuego-Patagonia (Southernmost South America): a spatial statistical approach. J Anthropol Archaeol 44:114–126. https://doi.org/10.1016/j.jaa.2016.09.006

Pallo MC, Borrazzo KB (2016) Miraflores rocks in Tierra del Fuego (southernmost South America): hunter-gatherer procurement, optimal pathway analysis and social interaction. J Archaeol Sci Rep 7:420–431. https://doi.org/10.1016/j.jasrep.2016.05.024

Pallo MC, Borrero LA (2015) ¿Intercambio o movilidad? Una evaluación sobre el uso de escalas de análisis espaciales y curvas de declinación en Patagonia centro-meridional (Argentina). Lat Am Antiq 26:287–303. https://doi.org/10.7183/1045-6635.26.3.287

Piana EL, Orquera LA (2007) Diferencias regionales y temporales en el litoral sudoccidental de Sudamérica. In: Morello F, Martinic M, Prieto A, Bahamonde G (eds) Arqueología de Fuego-Patagonia. Levantando piedras, desenterrando huesos y develando arcanos. CEQUA, Punta Arenas, pp 311–323

Piana EL, Orquera LA (2010) Shell midden formation at the Beagle Channel (Tierra del Fuego, Argentine). In: Calado D, Baldia M, Boulanger M (eds) Monumental Questions: Prehistoric Megaliths, Mounds and Enclosures. British Archaeological Reports International Series 2122, Archaeopress, Oxford, pp 263–273

Piana EL, Vázquez MM, Rua N (2004) Mischiuen I. Primeros resultados de una excavación de rescate en la costa norte del canal Beagle. In: Civalero T, Fernández P, Guráieb P (eds) Contra viento y marea. Arqueología de la Patagonia. INAPL, Buenos Aires, pp 815–832

Piana EL, Tessone A, Zangrando AF (2006) Contextos mortuorios en la región del canal Beagle... del hallazgo fortuito a la búsqueda sistemática. Magallania 34(1):87–101

Piana EL, Zangrando AF, Orquera LA (2012). Early occupations in Tierra del Fuego and the evidence from S layer in Imiwaia I site (Beagle Channel, Argentina). In: Miotti L, Salemme M, Flegenheimer N, Goebel T (eds) Current Research in the Pleistocene, Center for the Study of the First American. Southbound. Late Pleistocene Peopling of Latin America (Special edition). Texas A&M University, Center for the Study of the First Americans, pp 171–175

Pisano E (1977) Fitogeografía de Fuego-Patagonia Chilena. I. Comunidades vegetales entre las latitudes 52° y 56° S. Anal Inst Patag 8:121–250

Ponce JF, Borromei AM, Menounos B, Rabassa JO (2017) Late-Holocene and Little Ice Age palaeoenvironmental change inferred from pollen analysis, Isla de los Estados, Argentina. Quat Int 442:26–34. https://doi.org/10.1016/j.quaint.2016.04.016

Ponce JF, Rabassa J, Martínez O (2009) Morfometría y génesis de los fiordos de la Isla de los Estados, Tierra del Fuego. Rev Asoc Geol Argent 65:638–647. https://revista.geologica.org.ar/raga/article/view/865

Ponce JF, Borromei AM, Rabassa JO, Martinez O (2011) Late Quaternary palaeoenvironmental change in western Staaten Island (54.5° S, 64° W), Fuegian Archipelago. Quat Int 233(2):89–100. https://doi.org/10.1016/j.quaint.2010.10.011

Posth C, Nakatsuka N, Lazaridis I, Fehren-Schmitz L, Krause J, Reich D (2018) Reconstructing the deep population history of central and South America. Cell 175:1185–1197. https://doi.org/10.1016/j.cell.2018.10.027

Prates L, Politis GG, Pérez SI (2020) Rapid radiation of humans in South America after the last glacial maximum: a radiocarbon-based study. PLoS ONE 15(7):e0236023. https://doi.org/10.1371/journal.pone.0236023

Premoli A, Mathiasen P, Kitzberger T (2010) Southern-most Nothofagus trees enduring ice ages: genetic evidence and ecological niche retrodiction reveal high latitude (54S) glacial refugia. Palaeogeogr Palaeoclim Palaeoecol 298:247–256. https://doi.org/10.1016/j.palaeo.2010.09.030

Prieto A, Stern CR, Estévez J (2013) The peopling of the Fuego-Patagonian fjords by littoral hunter–gatherers after the mid-Holocene H1 eruption of Hudson Volcano. Quat Int 317:3–13. https://doi.org/10.1016/j.quaint.2013.06.024

Prieto A, Cárdenas P, Bahamonde G, Massone M (2004) Hallazgo de una fuente de materia prima en el chorrillo Miraflores, Tierra del Fuego, Chile. Magallania 32:229–232. https://magallania.cl/index.php/magallania/article/view/1213

Rabassa JO, Heusser CJ, Stuckenrath R (1986) New data on Holocene sea transgression in the Beagle Channel: Tierra del Fuego, Argentina. Quat S Am Antarct Penins 4:291–309

Rabassa JO, Heusser CJ, Rutter N (1990) Late Glacial and Holocene sea transgression in the Beagle Channel (Tierra del Fuego). Quat S Am Antarct Penins 7:335–360

Rabassa JO, Bujalesky G, Meglioli A, Coronato AMJ, Gordillo S, Roig C, Salemme M (1992) The Quaternary of Tierra del Fuego, Argentina: the status of our knowledge. Sver Geol Unders 81:249–256

Rabassa JO, Coronato AMJ, Bujalesky G, Roig C, Salemme M, Meglioli A, Heusser CJ, Gordillo S, Roig Juñet F, Borromei AM, Quattrocchio ME (2000) Quaternary of Tierra del Fuego, Southernmost South America: an updated review. Quat Int 68–71:217–240. https://doi.org/10.1016/S1040-6182(00)00046-X

Rabassa JO, Brandani A, Boninsegna J, Cobos D (1984) Cronología de la "Pequeña Edad de Hielo" en los glaciares Rio Manso y Castaño Overo, Cerro Tronador, Provincia de Río Negro. In: Abstracts of the IX Congreso Geológico Argentino, Actas 3, San Carlos de Bariloche, pp 624–639

Rabassa JO, Coronato AMJ, Roig C, Martínez O, Serrat D (2004) Un bosque sumergido en Bahía Sloggett, Tierra del Fuego, Argentina: evidencia de actividad neotectónica en el Holoceno Tardío. In: Blanco Chao R, López Bedoya J, Pérez Alberti A (eds) Procesos geomorfológicos y evolución costera. In: Actas de la II Reunión de Geomorfología Litoral. Universidad de Santiago de Compostela, Santiago de Compostela, pp 333–345

Ratto N (1991) Análisis funcional de las puntas de proyectil líticas de sitios del sudeste de la isla grande de Tierra del Fuego. Arqueología 1:151–178

Ratto N (1994) Funcionalidad versus adscripción cultural: cabezales líticos de la margen norte del estrecho de Magallanes. In: Lanata JL, Borrero LA (eds) Arqueología de cazadores-recolectores. Límites, casos y aperturas. Arqueología contemporánea 5. Edición Especial, pp 105–120

Renfrew C (1977) Alternative models for exchange and spatial distribution. In: Earle T, Ericson J (eds) Exchange Systems in Prehistory. Academic Press, New York, pp 71–90

Reyes O (2021) The settlement of the Chonos Archipelago, Western Patagonia. Springer, Cham, Chile

Robb J (2017) "Art" in archaeology and anthropology: an overview of the concept. Camb Archaeol J 27(4):587–597. https://doi.org/10.1017/S0959774317000725

Romano V, Lozano S, Fernández-López de Pablo J (2021) Reconstructing social networks of Late Glacial and Holocene hunter–gatherers to understand cultural evolution. Philos Trans R Soc B 377:20200318. https://doi.org/10.1098/rstb.2020.0318

Romero Villanueva G, Sepúlveda M, Cárcamo-Vega J, Cherkinsky A, De Porras ME, Barberena R (2024) Earliest directly dated rock art from Patagonia reveals socioecological resilience to mid-Holocene climate. Sci Adv 10(7). https://doi.org/10.1126/sciadv.adk4415

Sahlins M (1977) Economía de la Edad de Piedra. Akal, Madrid

Salemme M, Heusser C, Roig C, Coronato A, Rabassa JO (1995) Fire in Tierra del Fuego, southern South America. Charcoal particles from the Late Glacial/Early Holocene and their relationship

with the regional peopling. In: Abstracts of the 14th INQUA Congress, Berlin, 3–10 August 1995
Salemme M, Oría J, Turnes L, Santiago F (2014) Caracterización tecnomorfológica de dos conjuntos líticos del Holoceno medio en la Estepa Fueguina: La Arcillosa 2 y Río Chico 1. Rev Mus Antropol 7(1):39–54. https://doi.org/10.31048/1852.4826.v7.n1
San Román M, Prieto A (2004) (Dis)continuidad del uso de la obsidiana verde entre poblaciones de adaptación marítima del mar de Otway y estrecho de Magallanes. In: Civalero MT, Fernández P, Guráieb G (eds) Contra Viento y Marea. Arqueología de Patagonia. INAPL, Sociedad Argentina de Antropología, Buenos Aires, pp 149–166
San Román M, Reyes O, Torres Elgueta J, Morello F (2016) Archaeology of Maritime hunter-gatherers from Southernmost Patagonia, South America: discussing timing, changes and cultural traditions during the Holocene. In: Bjerck HB, Breivik H, Fretheim S, Piana E, Skar B, Tivoli A, Zangrando AF (eds) Marine Ventures. Archaeological Perspectives on Human-Sea Relations. Equinox, Sheffield, pp 157–174
San Roman M, Björck S, Nilsson B, Sjöström A, Borrero LA, Trejo V, Martínez I, McCulloch R, Huidobro C, Torres J, Mardones J, Lukoviek E, Morello F (2023) Early Maritime Peopling of the Pacific Tundra and the First Archaeological records from Brecknock Peninsula, Tierra del Fuego (54°S/72° W). In: Abstracts of the INQUA, Roma, 14–20 July 2023
San Román M (2013) Sitios arqueológicos de isla Englefield, mar de Otway: nuevas evidencias de discontinuidad cultural en el proceso de poblamiento marítimo de Patagonia meridional. In: Zangrando AF, Barberena R, Gil A, Neme G, Giardina L, Luna L, Otaola C, Paulides L, Salgán L, Tivoli AM (eds) Tendencias Teórico-Metodológicas y Casos de Estudio en la Arqueología de la Patagonia. Museo de Historia Natural de San Rafael, Mendoza, Sociedad de Antropología Argentina, INAPL, Buenos Aires, pp 523–534
Scheinsohn V (1997) Explotación de materias primas óseas en la Isla Grande de Tierra del Fuego. PhD Dissertation, Universidad de Buenos Aires
Scheinsohn V (2010) Hearts and Bones: bone raw material exploitation in Tierra del Fuego. BAR International Series 2094, Oxford
Scheinsohn V (2011) El trabajo del hueso en el fin del mundo: tecnología ósea en Bahía Valentín. In: Zangrando AF, Vázquez MM, Tessone A (eds) Los cazadores-recolectores del extremo oriental fueguino. Arqueología de Península Mitre e Isla de los Estados. Sociedad Argentina de Antropología, Buenos Aires, pp 271–286
Shennan S, Downey SS, Timpson A, Edinborough K, Colledge S, Kerig T, Manning K, Thomas MG (2013) Regional population collapse followed initial agriculture booms in mid-Holocene Europe. Nat Commun 4:2486. https://doi.org/10.1038/ncomms3486
Sierralta Navarro SG (2024) El ritmo de las piedras. Economía de la tecnología lítica en los archipiélagos occidentales de Patagonia. Un estudio desde Reloncaví y el norte de Chiloé. PhD Dissertation, Universidad Nacional del Centro de la Provincia de Buenos Aires
Steele J, Politis G (2009) AMS ^{14}C dating of early human occupation of southern South America. J Archaeol Sci 36:419–429. https://doi.org/10.1016/j.jas.2008.09.024
Stern C (2000) Fuentes de los artefactos de obsidiana en los sitios arqueológicos de las Cuevas de Pali AIke y Fell, y Cañadón La Leona, en Patagonia Austral. Anal Inst Patag Ser Cs Hum 28:251–263
Stern C (2008) Holocene tephrochronology record of large explosive eruptions in the southernmost Patagonian Andes. Bull Volcanol 70:435–454. https://doi.org/10.1007/s00445-007-0148-z
Stern C (2018) Obsidian sources and distribution in Patagonia, southernmost South America. Quat Int 468:190–205. https://doi.org/10.1016/j.quaint.2017.07.030
Stern C (2004) Obsidian in Southern Patagonia: review of the current information. In: Civalero MT, Fernández P, Guráieb G (eds) Contra Viento y Marea. Arqueología de Patagonia. INAPL, Sociedad Argentina de Antropología, Buenos Aires, pp 167–176

Timpson A, Colledge S, Crema E, Edinborough K, Kerig T, Manning K, Thomas MG, Shennan S (2014) Reconstructing regional population fluctuations in the European Neolithic using radiocarbon dates: a new case-study using an improved method. J Archaeol Sci 52:549–557. https://doi.org/10.1016/j.jas.2014.08.011
Tivoli AM (2013) Aprovechamiento de materias primas óseas de aves para la confección de punzones huecos en la región del canal Beagle. Intersecc Antropol 14(1):251–262
Tivoli AM, Zangrando AF (2011) Subsistence variations and landscape use among maritime hunter-gatherers. A zooarchaeological analysis from the Beagle Channel (Tierra del Fuego, Argentina). J Archaeol Sci 38:1148–1156. https://doi.org/10.1016/j.jas.2010.12.018
Tivoli AM, Montes A, Alunni DV, Fernández Ropero MC, Martinoli MP, Pérez SA, Pinto Vargas G, López R, Sáenz Samaniego R, Zangrando AF (2022) Arqueología de sectores insulares en el canal Beagle (Argentina). El caso del archipiélago de las islas Bridges e islas adyacentes. Arqueología 28(1):9976. https://doi.org/10.34096/arqueologia.t28.n1.9976
Torrence R (1989) Tools as optimal solutions. In: Torrence R (ed) Time, energy and stone tools. Cambridge University Press, Cambridge, pp 1–6
Tuhkanen S (1992) The climate of Tierra del Fuego from a vegetation geographical point of view and its ecoclimatic counterparts elsewhere. Acta Bot Fenn 145:1–64
Unkel I, Fernandez M, Björck S, Kjung K, Wolfarth B (2010) Records of environmental changes during the Holocene from Isla de los Estados (54.4°S), southeastern Tierra del Fuego. Glob Planet Change 74:99–113. https://doi.org/10.1016/j.gloplacha.2010.07.003
Vázquez MM, Zangrando AF, Tessone A, Ceraso A (2011) Investigaciones arqueológicas recientes en la costa suroriental de Península Mitre. In: Zangrando AF, Vázquez MM, Tessone A. (eds) Los cazadores-recolectores del extremo oriental fueguino. Arqueología de Península Mitre e Isla de los Estados. Sociedad Argentina de Antropología, Buenos Aires, pp 203–203
Vázquez MM (2024) Movilidad y Subsistencia de los cazadores-recolectores de Península Mitre, extremo sudeste de Tierra del Fuego. PhD Dissertation, Universidad de Buenos Aires
Vidal H (2011[1985]) El conocimiento de las culturas aborígenes del sudeste de la Isla Grande hasta 1983. In: Zangrando AF, Vázquez MM, Tessone A. (eds) Los cazadores-recolectores del extremo oriental fueguino. Arqueología de Península Mitre e Isla de los Estados. Sociedad Argentina de Antropología, Buenos Aires, pp 65–113
Villalba R (1990) Climatic fluctuations in northern Patagonia during the last 1000 years as inferred from tree-rings records. Quat Res 34:346–360. https://doi.org/10.1016/0033-5894(90)90046-N
Waldmann N, Ariztegui D, Anselmetti FS, Austin JA, Moy CM, Stern C, Recasens C, Dunbar RB (2010) Holocene climatic fluctuations and positioning of the Southern Hemisphere Westerlies in Tierra del Fuego (54°S), Patagonia. J Quat Sci 25:1063–1075. https://doi.org/10.1002/jqs.1263
Waldmann N, Borromei AM, Recacens C, Olivera D, Martinez MA, Maidana NI, Ariztegui D, Austin JA, Anselmetti FS, Moy CM (2014) Integrated reconstruction of Holocene millennial-scale environmental changes in Tierra del Fuego, southernmost South America. Palaeogeogr Palaeoclim Palaeoecol 399:294–309. https://doi.org/10.1016/j.palaeo.2014.01.023
Williams A (2012) The use of summed radiocarbon probability distributions in archaeology: a review of methods. J Archaeol Sci 39:578–589. https://doi.org/10.1016/j.jas.2011.07.014
Yesner D (1990) Fuegian and other hunter-gatherers of Subantarctic region: “cultural devolution” reconsidered. In: Meehan BJ, White N (eds) Hunter-gatherer demography: past and present. University of Sydney, Sidney, pp 1–11
Zangrando AF (2010) Coastal archaeology and hunter-gatherers in the south-eastern of Tierra del Fuego. J Isl Coast Archaeol 5(2):288–291. https://doi.org/10.1080/15564894.2010.487360
Zangrando AF (2018) Poblamiento temprano y arqueología de costas en Patagonia y Tierra del Fuego: vacío de información, preconceptos y perspectivas. Intersecc Antropol 19:63–97
Zangrando AF, Ponce JF, Martinoli MP, Montes A, Piana EL, Vanella F (2016) Palaeogeographic changes drove prehistoric fishing practices in the Cambaceres Bay (Tierra del Fuego, Argentina) during the middle and late Holocene. Environ Archaeol 21(2):182–192. https://doi.org/10.1080/14614103.2015.1130888

Zangrando AF, Bjerck HB, Piana EL, Breivik HM, Tivoli AM, Negre J (2018) Spatial patterning and occupation dynamics during the early Holocene in an archaeological site from the south coast of Tierra del Fuego: Binushmuka I. Estud Atacameños 60:31–49

Zangrando AF, Tessone A, Vázquez M (2009) El uso de espacios marginales en el Archipiélago Fueguino: implicaciones de la evidencia arqueológica de Bahía Valentín. In: Salemme M, Santiago F, Álvarez M, Piana E, Vázquez M, Mansur E (eds) Arqueología de Patagonia: una mirada desde el último confín. Utopías, Ushuaia, pp 47–62

Zangrando AF, Tivoli AM, Ponce JF, Alunni DA, Fernández Ropero MC, Martinoli MP (2021a) Colonización humana de Tierra del Fuego: contextos arqueológicos pre-transgresivos en la costa norte del Canal Beagle. Chungara 54(3):557–577. https://doi.org/10.4067/S0717-73562022005000401

Zangrando AF, Tivoli AM, Fernández Ropero C, Pinto Vargas G (2021b) Contextos arqueológicos en paleocostas: un nuevo caso de estudio en Bahía Moat (Tierra del Fuego). Paper presented at the XI Jornadas de Arqueología de la Patagonia 2021, Universidad Austral de Chile, Puerto Montt, 29 November–3 December 2021

Zangrando, AF, Martinoli MP, Tessone A, Vázquez MM, Alunni DV, Franch Bach A, Hernández Herrero O, Fernández Ropero MC, Pinto Vargas G, Tivoli AM (2024) El uso humano de espacios externos en el archipiélago de Tierra del Fuego durante el Holoceno: el caso de Isla de los Estados. Bol Soc Chil Arqueol 57. https://doi.org/10.56575/BSCHA.05700240843

Zangrando AF (2009) Historia evolutiva y subsistencia de cazadores-recolectores marítimos de Tierra del Fuego. Sociedad Argentina de Antropología, Buenos Aires

Zangrando AF (2022) Colonización humana de los canales fueguinos: revisión y perspectivas. Intersecc Antropol 23(1):153–166. https://doi.org/10.37176/iea.23.1.2022.783

Chapter 3
Human Subsistence Strategies on the Southern Coast of the Fuegian Andes: A Comparative Analysis of Vertebrate Faunal Exploitation During the Holocene

Angélica M. Tivoli, Daniela V. Alunni, María Paz Martinoli, and Martín M. Vázquez

Abstract This chapter assesses the use of faunal resources across broad temporal and spatial scales through the integration of zooarchaeological analyses conducted on assemblages distributed along the southern coast of Tierra del Fuego; from the northern shore of the Beagle Channel—beginning at its boundary with Chile—to Isla de los Estados, dating from approximately 7300 cal BP onward. The four taxonomic categories of vertebrates that are recurrent in the archaeological record are analyzed here: pinnipeds, guanacos, birds, and fish. These categories are represented in most of the studied sites and are commonly reported in terms of their Number of Identified Specimens (NISP). In general terms, the zooarchaeological analysis of the southern coastal region reveals a strong reliance on marine resource consumption. The southern coast of Tierra del Fuego is a region characterized by a relatively homogeneous resource base, however, some differences have been observed regarding the relative abundance of various taxa along different sectors of the southern coast of Tierra del Fuego. Morover, this study identifies regional patterns of change and continuity in subsistence strategies throughout the Holocenoe—trends that likely reflect broader processes of population expansion and settlement, in which the occupation of different coastal spaces occurred at distintc tempos. Approaching the southern

A. M. Tivoli (✉) · D. V. Alunni · M. P. Martinoli · M. M. Vázquez
CADIC-CONICET, Ushuaia, TDFeIAS, Argentina
e-mail: tivoli.angelica@conicet.gov.ar

D. V. Alunni
e-mail: daniela.alunni@conicet.gov.ar

M. P. Martinoli
e-mail: mpmartinoli@conicet.gov.ar

M. M. Vázquez
e-mail: vazquezmartin68@gmail.com

M. M. Vázquez
UNTDF, Ushuaia, TDFeIAS, Argentina

A. F. J. Zangrando et al. (eds.), *Archaeology of the South Coast of Tierra del Fuego*, The Latin American Studies Book Series, https://doi.org/10.1007/978-3-032-09487-2_3

Fuegian coast as an integrated system has enabled the identification and characterization of temporal trends in faunal abundance and exploitation strategies, within the broader context of human use of coastal environments across different stages of Fuegian littoral occupation.

3.1 Introduction

Zooarchaeological studies carried out in different areas of the southern coast of Tierra del Fuego (Argentina) have resulted in a substantial body of research which reveals a variable consumption of marine and terrestrial resources, as well as differences in their exploitation patterns throughout the Holocene (Vidal 1985; Orquera and Piana 1999a; Zangrando 2009a; Zangrando et al. 2009; Tivoli 2010a; Tivoli and Zangrando 2011; Horwitz and Weissel 2011; Vázquez et al. 2011; Alunni 2018a, 2018b; Martinoli 2018, 2019; Vázquez 2024; among others).

While some previous studies have undertaken regional syntheses (Santiago and Vázquez 2012; Zangrando and Tivoli 2015), the exploitation of the four most abundant vertebrate categories in the archaeological assemblages—pinnipeds, guanacos, fish, and birds—has not yet been evaluated comprehensively and diachronically for the southern coast of Tierra del Fuego. This chapter assesses the use of faunal resources across broad temporal and spatial scales through the integration of zooarchaeological analyses conducted on assemblages distributed from the northern shore of the Beagle Channel—beginning at its boundary with Chile—to Isla de los Estados. This approach offers a novel perspective on the formation of the archaeological record (Lyman 2003; Bailey 2008), as it enables the exploration of subsistence strategies within the broader context of regional human settlement processes across diverse environmental settings (Horwitz 1990; Legoupil and Fontugne 1997; Scheinsohn 1997, 2010; Orquera and Piana 2006; Zangrando 2009a; Zangrando et al. 2009; Horwitz and Weissel 2011; Vázquez 2024; Zangrando et al. 2024a).

3.2 Environmental Scenario of the Southern Coast of Tierra del Fuego

The southernmost tip of South America features an archipelago of islands of varying sizes that form an intricate network of channels that extends from the southern coast of the Strait of Magellan to Cape Horn, between approximately 52° and 56° South latitude. These coastal landscapes are characterized by mountainous terrain with slopes covered by dense forests of the genus *Nothofagus*, interspersed with peat bogs and grassland areas (Iturraspe and Schroeder 1999; Collado 2001; Coronato et al. 2007; Coronato 2014).

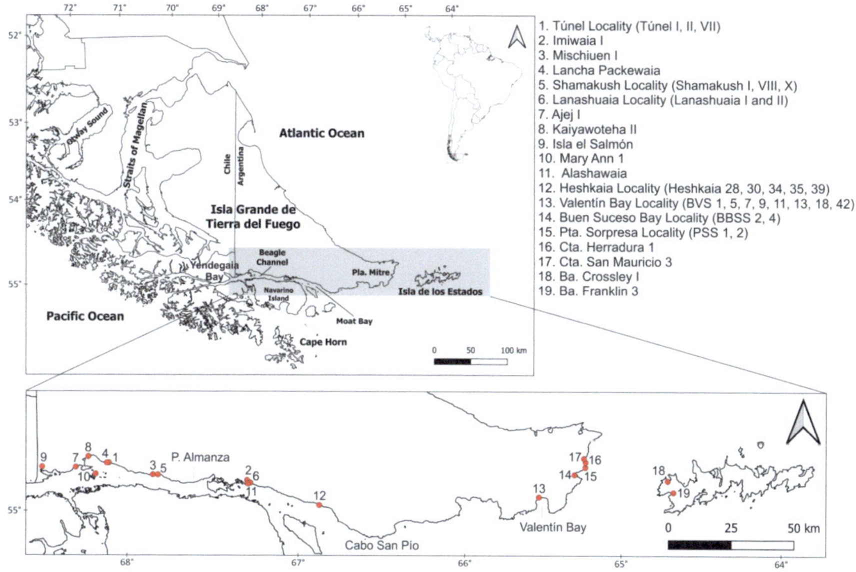

Fig. 3.1 Map of the study region, showing the main geographical references and the location of the archaeological sites studied

South of Tierra del Fuego Island lies the Beagle Channel or Onashaga[1] (Fig. 3.1), a tectonic valley that was occupied during the last glaciation by the Beagle paleoglacier and subsequently flooded by marine waters (Gordillo et al. 1992; Bujalesky 2011; Borromei et al. 2014). This channel is located between 54° and 55° south latitude, runs from west to east, and connects the Pacific and Atlantic Oceans (Fig. 3.1). The central sector, between Yendegaia Bay and Puerto Almanza, is characterized by mountainous terrain and predominantly steep coasts, interspersed with small beaches and relatively long stretches of low coastline. In the eastern sector of the channel, from Puerto Almanza to Cabo San Pío, the height of the mountainous backdrop decreases, and the landscape is marked by prominent moraine arcs and fields of drumlins, forming gently rolling hills, low-relief terrain, and broad, gently sloping beaches (Orquera and Piana 1999b; Rabassa et al. 2000).

At the southeastern tip of Isla Grande lies the Península Mitre, whose southern coasts are predominantly rocky and steep, interspersed with wide bays: Sloggett, Aguirre, Valentín, and Buen Suceso (Vázquez et al. 2011). East of the peninsula, across the Le Maire Strait (approximately 30 km; Ponce et al. 2011), is located Isla de los Estados, characterized by a rugged relief, with a series of cliffs interspersed with small beaches (Caminos and Nullo 1979).

The general climate is temperate-cold oceanic, influenced by prevailing westerly and southwesterly winds (Iturraspe and Schroeder 1999; Rabassa et al. 2000; Coronato 2014). The average annual temperature has been estimated at 5.3 °C, and rainfall

[1] Onashaga: name in the language of the Yagan First Nation.

in the central Beagle Channel ranges between 500 and 600 mm per year (Iturraspe and Schroeder 1999; Coronato 2014), while in Moat it reaches values of 780 mm and 1000 mm in the eastern end of Isla Grande (Orquera and Piana 1999a; Coronato 2014). On Isla de los Estados, rainfall is very frequent and averages 1450 mm per year (Dudley and Crow 1983; Ponce and Fernández 2014). There are also differences in the hydrological regimes. On the northern shore of the central Beagle Channel, the drainage network consists of short, moderate-flow streams that descend steep slopes. Mountain basins predominate, fed by the seasonal accumulation of snow in upper valleys and by numerous small glaciers (Zangrando 2009a; Zangrando et al. 2014). By contrast, to the east, in the Bahía Moat sector and on the Península Mitre, the river systems correspond to peat basins. Glaciers are absent in this area, and snow accumulation is of limited persistence. Peat bogs predominate in this environment, forming extensive subsurface reservoirs and small ponds (Roig and Collado 2004). On Isla de los Estados, watercourses are short and permanent due to abundant rainfall and the high water content of the soil (Ponce 2009).

3.3 Resource Availability

The southern coast of Tierra del Fuego offers a wide variety of resources (Table 3.1), although not all of them may have been attractive for human subsistence. Zooarchaeological evidence suggests hunter-gatherers utilized different spaces within the region (Zangrando 2009a; Tivoli and Zangrando 2011; Vázquez et al. 2011; Horwitz and Weissel 2011). Accordingly, we revisit the ecological zonation proposed in earlier studies (Zangrando 2009a; Tivoli and Zangrando 2011), which define three main ecozones: terrestrial, littoral, and marine (offshore).

The terrestrial zone is characterized by the presence of birds and a few mammal species, while the abundance and diversity of freshwater fish are low (Zangrando 2009a). Among the avifauna, several species of raptors are common in this environment, including Chimango Caracara (*Phalcoboenus chimango*), small eagles, falcons, and the Andean Condor (*Vultur gryphus*). Scavenger birds such as the Crested Caracara (*Caracara plancus*) and the Turkey Vulture (*Cathartes aura*) are also present, along with the Austral Parakeet (*Enicognathus ferrugineus*), the Magellanic Woodpecker (*Campephilus magellanicus*), various genera and species of Anatidae such as the Upland Goose (*Chloephaga picta*), and numerous passerines (Couve and Vidal 2003; Narosky and Izurieta 2003).

Mammals include the Fuegian fox (*Lycalopex culpaeus lycoides*), small rodents, and guanacos (*Lama guanicoe*). This camelid was the only native terrestrial mammal of significant economic importance for the subsistence of the region's early human populations (Orquera and Piana 1999a; Vázquez 2015; Alunni 2018a, 2018b; among others). In the forested environments of the southern part of the island, guanaco availability may have varied seasonally according to altitudinal fluctuations in the mountainous zones (Flores et al. 2018). Harsh winter conditions at higher elevations typically lead to downward migratory movements, resulting in increased guanaco

Table 3.1 Distribution of current native animal resources relevant for human subsistence across terrestrial and marine ecozones (modified from Tivoli and Zangrando 2011)

Ecozones	Mammals	Birds	Fish
Terrestrial	**Guanacos** (Camelidae: *Lama guanicoe*) **Fuegian foxes** (Canidae: *Lycalopex culpaeus lycoides*)	**Falcons and Caracaras** (Falconidae: *Phalcoboenus chimango*, *Caracara plancus*, etc.) **Scavengers** (Cathartidae: *Cathartes aura*, *Vultur gryphus*) **Eagles, hawks** (Accipitridae) **Herons** (Ardeidae: *Nycticorax nycticorax*) **Woodpeckers** (Picidae: *Campephilus magellanicus*) **Parakeet** (Psittacidae: *Enicognathus ferrugineus*) **Ducks, geese** (Anatidae: *Chloephaga* sp., *Lophonetta specularioides*, etc.) **Passerines** (Emberizidae, Furnariidae, etc.)	**Puyen** (Galaxiidae: *Galaxia maculatus*) in rivers and lagoons

(continued)

Table 3.1 (continued)

Ecozones	Mammals	Birds	Fish
Marine (near-shore)	**Pinnipeds** (Otariidae: *Arctocephalus australis*, *Otaria flavescens*) **Otters** (Mustelidae: *Lontra felina*, *Lontra provocax*) **Cetaceans** (beachings): **Odontoceti** Dolphins (Delphinidae: *Orcinus orca*, *Globicephala melas*, *Lagenorhynchus obscurus*, etc.)/ Whales (Ziphiidae: *Tasmacetus shepherdi*, *Berardius arnuxii*, *Mesoplodon hectori* and occasionally Physeteridae: *Physeter macrocephalus*)/Porpoises (Phocoenidae: *Phocoena spinipinnis*, *Phocoena dioptrica*) **Mysticeti** Baleen whales (*Megaptera novaeangliae*, *Balaenoptera musculus*, *Caperea marginata*, *Eubalaena australis*, etc.)	**Penguins** (Spheniscidae: *Spheniscus magellanicus*, *Eudyptes chrysocome*) **Cormorants** (Phalacrocoracidae: *Leucocarbo* sp.) **Ducks, geese, swans** (Anatidae: *Tachyeres* sp., *Chloephaga* sp., *Lophonetta specularioides*, etc.) **Gulls** (Laridae: *Larus* sp.) **Skuas** (Stercorariidae: *Catharacta* sp.) **Petrels, shearwaters and fulmars** (Procellariidae) **Albatrosses** (Diomedeidae)	**Notothenioids** (Nototheniidae: *Patagonotothen* sp.; Eleginopidae: *Eleginops maclovinus*, etc.) **Eelpouts** (Zoarcidae: *Austrolycus* sp.) **Channel Bull Blenny** (Bovichthyidae: *Cottoperca gobio*) **Puyen** (Galaxiidae: *Galaxia maculatus*) **Patagonian grenadier** (beachings) (Merluciidae: *Macruronus magellanicus*) **Sardines** (Clupeidae: *Sprattus fuegensis*)
Marine (offshore)	**Odontoceti**: Dolphins/Whales **Mysticeti**: Baleen whales	**Petrels, shearwaters and fulmars** (Procellariidae: *Macronectes giganteus*, *Procellaria* sp., *Daption capense*, *Ardenna* sp., *Puffinus puffinus*, *Fulmarus glacialoides*, etc.) **Albatrosses** (Diomedeidae)	**Barracouta** (Gempylidae: *Thyrsites atun*) **Hakes** (Merluciidae: *Merluccius australis*) **Tadpole codling** (Moridae: *Salilota australis*) **Pink cusk-eel** (Ophidiidae: *Genypterus blacodes*)

presence along the coast during this season—particularly in the central sector of the Beagle Channel, where the mountain range reaches its greatest height. Conversely, their abundance would decrease in summer as they move inland due to the greater forage availability (Moraga et al. 2015). The lower mountain range altitude in the eastern Beagle Channel and Península Mitre facilitates coastal-inland movement, resulting in less seasonal guanaco variation, particularly on the Península Mitre (Lanata 1996; Orquera and Piana 1999a). Finally, no guanacos are recorded in Isla de los Estados.

The littoral ecozone, corresponding to the near-shore marine environment, forms the interface between the marine and terrestrial systems. Coastal margins provide the largest biomass of resources for human subsistence in the region and offer favorable conditions in terms of accessibility and predictability (Zangrando 2009a). This ecozone is primarily inhabited by two pinniped species from the Otariidae family: South American sea lions (*Otaria flavescens*) and South American fur seals (*Arctocephalus australis*). Their current distribution varies according to the type of coastline chosen to establish their rookeries: the former prefer sheltered coasts and can be found both on islands and islets within the Beagle Channel (e.g., Becasses Islands) and in more open areas (e.g., Península Mitre and Isla de los Estados), whereas South American fur seals are more commonly located in exposed sectors such as Isla de los Estados or Cape Horn (Sielfeld 1983; Schiavini 1990; Schiavini and Raya Rey 2001; Milano et al. 2020a, 2020b). Both species are polygynous, with an annual cycle divided into a reproductive stage in summer and a non-reproductive phase when individuals visit coastal areas to form resting haul-outs. Adult males and immature individuals typically gather at separate sites and spend more time in the water foraging. Females repeatedly return to breeding areas to nurse their pups for a one-year period (Bowen et al. 2009). Regarding size, *A. australis* reaches the following dimensions: pups between 3.5 m and 5.5 kg; adult males: 1.9 m and 159 kg; adult females: 1.4 m and 48.5 kg (Vaz Ferreira and Vallejo 1980). The average sizes of *O. flavescens* are: pups between 10.2 and 14.8 kg; adult males: 2.56 m and 300–340 kg (Vaz Ferreira 1981); adult females: 2 m and 144 kg (Schiavini 1990).

The littoral ecozone also supports significant avian diversity, including penguins, cormorants, anatids, gulls, and other species (Raya Rey and Schiavini 2000; Schiavini and Raya Rey 2001). Some of these birds are gregarious, such as cormorants, penguins, and gulls. Among the cormorants, Imperial Cormorants (*Leucocarbo atriceps*) form large colonies, often on islands or islets associated with sea lion haul-outs (Schiavini and Yorio 1995; Schiavini and Raya Rey 2001). Currently, Magellanic Penguins (*Spheniscus magellanicus*) nest within the Beagle Channel from late spring to the end of summer. Gentoo Penguins (*Pygoscelis papua*) also breed in the area, and King Penguins (*Aptenodytes patagonicus*) can occasionally be observed in the eastern sector. In particular, Isla de los Estados hosts various bird colonies, including those of cormorants, Magellanic and Rockhopper Penguins (*Eudyptes chrysocome*), and Southern Giant Petrels (*Macronectes giganteus*) (Schiavini and Raya Rey 2001). Waterfowl include several species of geese, steamer ducks, and other smaller ducks. Shallow coastal waters are rich in fish, predominantly small Nototheniidae species like doraditos and lorchos (Lloris and Rucabado 1991). The Patagonian Blennie

(*Eleginops maclovinus*) is the species that reaches the greatest length, weight, and biomass and is most abundant between October and April (López et al. 1996). Additionally, in this ecozone there are also beds of different species of molluscs and crustaceans, such as the Southern King Crab (*Lithodes santolla*) and the False King Crab (*Paralomis granulosa*) (Orquera and Piana 1999a; Lovrich 2014).

Lastly, within the marine ecozone, various species of pelagic (offshore) birds are notable, including petrels, shearwaters, fulmars, and albatrosses. These taxa exhibit reproductive behaviors involving nesting on both coastal and outer islands of the Fuegian archipelago, as in the case of the Southern Giant Petrel (*Macronectes giganteus*) colony on Isla de los Estados. Some species also regularly enter the Beagle Channel (Schiavini and Yorio 1995; Raya Rey and Schiavini 2000; Schiavini and Raya Rey 2001). Indeed, external areas such as the Península Mitre and Isla de los Estados are zones of high marine productivity, resulting from the confluence of the Malvinas Current and the Antarctic Circumpolar Current (Acha et al. 2004). The ichthyofaunal community in deep waters displays distinctly different structural traits compared to those found in coastal areas (Lloris and Rucabado 1991). Both large-bodied species, such as Southern Hake (*Merluccius* sp.), and smaller pelagic taxa, such as sardines, are present. While certain species exhibit strictly pelagic behavior—e.g., Snoek (*Thyrsites atun*)—others may occasionally approach coastal areas; Fuegian Sardines (*Sprattus fuegensis*) and hake, for instance, are known to produce large-scale strandings (Lloris and Rucabado 1991). Meanwhile, several species of pelagic cetaceans are known to frequent the southern coast of Tierra del Fuego. These include a variety of dolphins and baleen whales, two species of porpoises, and, occasionally, Sperm Whales (*Physeter macrocephalus*). These marine mammals may occasionally strand on the coast (Goodall 1978; Goodall et al. 2008).

3.4 Regional Archaeological Background

Human occupations along the southern coast of Tierra del Fuego have been documented since the Early Holocene—around 8600 cal yr BP (Piana et al. 2012; Zangrando et al. 2018, 2021a, 2021b; Tivoli et al. 2022). However, only a few sites with this chronology have been identified within the Beagle Channel, and in these cases, bone remains are scarce, highly fragmented, and often calcined (Piana et al. 2012; Zangrando et al. 2021a, 2021b). Around 7300 cal yr BP, a series of changes becomes evident: (1) multicomponent shell midden deposits make the archaeological record more visible; (2) evidence of human occupation begins to spread across nearly all areas of the Fuegian archipelago, although at different rates; and (3) the composition and formation processes of archaeological sites allow for the preservation of abundant technological and faunal remains (Ortiz Troncoso 1975; Orquera and Piana 1983, 1988, 1989–90, 1999a, 2009; Vidal 1985; Legoupil 1993–94; Legoupil and Fontugne 1997; Ocampo and Rivas 2000, 2002; Zangrando 2009a, 2018; Orquera et al. 2011; Zangrando et al. 2009, 2011, 2021b; Vázquez 2024; among others). For these reasons, this study focuses on assemblages dated to later than 7300 cal yr BP (Table 3.2), as they provide a higher degree of taxonomic identification and enable the establishment of statistically significant trends regarding subsistence strategies.

As previously noted, the various sectors comprising the southern coast of Tierra del Fuego have been subject to differing research trajectories and intensities (Vázquez and Prieto 2014). These studies have identified spatial differences in subsistence strategies (Orquera and Piana 1999a; Zangrando et al. 2009; Zangrando et al. 2010).

The research carried out by Orquera and Piana on the northern coast of the Beagle Channel led them to propose a model of adaptation to marine littoral life (Orquera and Piana 1983, 1988, 1999a, 2006, 2009; Piana 1984). At the core of this model is the notion that subsistence was primarily based on the hunting of adult and subadult males of the South American fur seal (*Arctocephalus australis*). This prey would have been captured mainly in aquatic settings during the winter season, implying the development of specific technologies such as watercraft and harpoons with detachable points (Orquera and Piana 1983, 1988, 1999a, 2006, 2009; Piana 1984; Schiavini 1990, 1993). The authors also argue that the relatively homogeneous spatial distribution of faunal resources along the coast would have allowed for similar access from different locations. This had two implications (Orquera and Piana 1999a): (1) highly mobile societies and (2) adaptive stability and equilibrium between Beagle Channel populations and their environment for more than six millennia—interrupted only by contact with European populations. However, the subsequent incorporation of new archaeological contexts with different characteristics and chronologies (Piana et al. 2000, 2004, 2007a; Orquera and Piana 2005, 2009; Piana and Canale 1993–94; Piana and Vázquez 2008; Zangrando 2009a, 2009b; Tivoli 2010a, 2010b; Álvarez et al. 2013) enabled the identification of temporal variations that suggest a reorientation of subsistence activities throughout the Holocene (Zangrando 2009a, 2009b; Tivoli 2010a, 2010b; Tivoli and Zangrando 2011). In general terms, around 7000 cal yr BP, the archaeological record shows a predominance of pinniped consumption and a marked specialization in the exploitation techniques for these resources (Orquera and Piana 1983, 1988, 1999a, 2006, 2009; Piana 1984; Zangrando 2009a). Later, around 5500 cal yr BP, a decrease in the relative frequency of pinniped remains is observed in the Beagle Channel, along with a relative increase in other resources, particularly guanacos and birds (Zangrando 2009a, 2009b; Tivoli and Zangrando 2011). Around 1300 cal yr BP, another turning point occurred, marked by an increase in the representation of birds and fish, as well as the more active incorporation of offshore areas into the organization of subsistence activities. This has been interpreted as a process of intensification in fishing practices and an expansion of the foraging range for birds and fish (Zangrando 2009a, 2009b; Tivoli 2010a, 2010b; Tivoli and Zangrando 2011). Research on technology and ornamentation also pointed to the existence of temporal changes that accompany these trends, as well as modifications in the zooarchaeological record (Orquera and Piana 1995, 1999a; Scheinsohn 1997, 2010; Álvarez 2003, 2005, 2009a, 2009b; Fiore 2006, 2011, 2012; Piana and Orquera 2007).

More recent studies carried out in Moat Bay—the eastern sector of the Beagle Channel—indicate that the archaeofaunal assemblages from the Late Holocene exhibit certain contrasts with those previously analyzed in other areas of the channel.

Table 3.2 Archaeological assemblages: chronology, results of NISP, NISP%, and references

Area	Assemblages	Chronology (radiocarbon years BP)	Calibrated years BP (Median probability)	Calibrated years BP: 95.4 (2 sigma)	GUANACOS		PINNIPEDS		BIRDS		FISH		TOTAL NISP	References
					NISP	NISP%	NISP	NISP%	NISP	NISP%	NISP	NISP%		
CENTRAL BEAGLE CHANNEL	Túnel I (2 Comp) Fase 1	6460 ± 110	7335	7156–7518	149	6.01	2223	89.60	70	2.82	39	1.57	2481	Orquera and Piana (1999a), Zangrando (2009a)
	Túnel I (2 Comp) Fase 2	–	–	–	305	3.98	6645	86.69	252	3.29	463	6.04	7665	
	Túnel I (2 Comp) Fase 3	5950 ± 170	6749	6392–7164	191	3.64	3288	62.58	185	3.52	1590	30.26	5254	
	Túnel I (2 Comp) Fase 4	–	–	–	752	5.70	11,346	86.00	901	6.83	194	1.47	13,193	
	Túnel I (2 Comp) Fase 5	5850 ± 70	6615	6442–6788	254	3.83	5067	76.43	450	6.79	859	12.96	6630	
	Túnel I (2 Comp) Fase 6	5050 ± 520	5719	4416–6944	642	10.18	4764	75.52	515	8.16	387	6.14	6308	
	Túnel I (2 Comp) Fase 7	–	–	–	830	12.58	4434	67.21	305	4.62	1028	15.58	6597	
	Túnel I (2 Comp) Fase 8	–	–	–	618	12.49	3398	68.69	527	10.65	404	8.17	4947	

(continued)

Table 3.2 (continued)

Area	Assemblages	Chronology (radiocarbon years BP)	Calibrated years BP (Median probability)	Calibrated years BP: 95.4 (2 sigma)	GUANACOS		PINNIPEDS		BIRDS		FISH		TOTAL NISP	References
					NISP	NISP%	NISP	NISP%	NISP	NISP%	NISP	NISP%		
	Mischiuen (F)	4430 ± 180	5022	4521–5476	1130	32.50	544	15.65	1156	33.25	647	18.61	3477	Piana et al. (2004), Tivoli and Zangrando (2011)
	Túnel I (3 Comp)	4300 ± 80	4822	4527–5047	741	36.79	313	15.54	800	39.72	160	7.94	2014	Orquera and Piana (1999a), Zangrando (2009a), Martinoli (2018)
	Lancha Packewaia (X)	4020 ± 70	4451	4233–4648	1646	44.67	1429	38.78	608	16.50	2	0.05	3685	Orquera and Piana (1993–94), Zangrando (2009a)
	Túnel I (4 Comp)	2660 ± 100 2690 ± 80	2724 2773	2408–2950 2666–2962	1225	43.83	516	18.46	960	34.35	94	3.36	2795	Orquera and Piana (1999a), Zangrando (2009a), Martinoli (2018)

(continued)

Table 3.2 (continued)

Area	Assemblages	Chronology (radiocarbon years BP)	Calibrated years BP (Median probability)	Calibrated years BP: 95.4 (2 sigma)	GUANACOS		PINNIPEDS		BIRDS		FISH		TOTAL NISP	References
					NISP	NISP%	NISP	NISP%	NISP	NISP%	NISP	NISP%		
	Túnel I (5 Comp)	1920 ± 80 1990 ± 110	1814 1,895	1692–2001 1692–2137	318	28.57	707	63.52	64	5.75	24	2.16	1113	Orquera and Piana (1999a), Zangrando (2009a)
	Isla el Salmón	1560 ± 90 1765 ± 25 1820 ± 120	1418 1646 1699	1272–1591 1576–1701 1416–1933	50	78.13	2	3.13	12	18.75	0	0	64	Figuerero Torres and Mengoni Goñalons (1986)
	Shamakush I (C, D, E, F)	890 ± 100 940 ± 110 1022 ± 100 1220 ± 110 1927 ± 120	773 817 877 1091 1821	637–932 652–987 715–1066 904–1303 1558–2095	1332	18.62	169	2.36	231	3.23	5422	75.79	7154	Orquera and Piana (1996), Zangrando (2009a)
	Shamakush X	500 ± 100 1480 ± 100	485 1345	303–571 1176–1541	9	1.59	9	1.59	173	30.62	374	66.19	565	Zangrando (2009a), Tivoli (2010a, 2010b)
	Lancha Packewaia (B, C, D)	280 ± 65 1080 ± 100 1120 ± 50 1590 ± 50	289 942 984 1,440	239–468 737–1116 903–1073 1346–1537	415	13.91	2082	69.77	487	16.32	0	0	2984	Orquera and Piana (1993–94)

(continued)

Table 3.2 (continued)

Area	Assemblages	Chronology (radiocarbon years BP)	Calibrated years BP (Median probability)	Calibrated years BP: 95.4 (2 sigma)	GUANACOS		PINNIPEDS		BIRDS		FISH		TOTAL NISP	References
					NISP	NISP%	NISP	NISP%	NISP	NISP%	NISP	NISP%		
	Ajej I	1270 ± 180 1400 ± 90	1130 1257	758–1431 1058–1426	10	0.57	1011	57.51	702	39.93	35	1.99	1758	Piana et al. (2008), Mameli and Estévez Escalera (2004), Martinoli (2018)
	Shamakush VIII	1490 ± 90 730 ± 55	1353 635	1178–1535 554–686	176	10.04	145	8.27	1432	81.69	0	0	1753	Piana and Vázquez (2008)
	Túnel II	1120 ± 90 1140 ± 90	989 1012	791–1178 878–1178	27	1.46	384	20.79	1436	77.75	0	0	1847	Piana and Canale (1993–94), Tivoli (2010a, 2010b)
	Mischiuen (C)	890 ± 90 1060 ± 85	771 917	652–930 732–1074	673	14.95	1946	43.23	1882	41.81	0	0	4501	Piana et al. (2004), Tivoli (2010a, 2010b)
	Mary Ann 1	1030 ± 40	868	797–874	12	0.41	94	3.20	578	19.71	2249	76.68	2933	Tivoli et al. (2022)

(continued)

Table 3.2 (continued)

Area	Assemblages	Chronology (radiocarbon years BP)	Calibrated years BP (Median probability)	Calibrated years BP: 95.4 (2 sigma)	GUANACOS		PINNIPEDS		BIRDS		FISH		TOTAL NISP	References
					NISP	NISP%	NISP	NISP%	NISP	NISP%	NISP	NISP%		
	Kaiyawoteha II (L, M, Ñ, T)	726 ± 41	633	621–680	45	0.23	7674	39.39	11,064	56.79	698	3.58	19,481	Piana et al. (2007b), Vázquez et al. (2018)
	Túnel I (6 Comp)	450 ± 60 650 ± 80	633	621–680	10	2.58	113	29.12	32	8.25	233	60.05	388	Orquera and Piana (1999a), Zangrando (2009a)
	Túnel VII	100 ± 90	–	–	117	0.51	8953	39.27	13,727	60.21	0	0	22,797	Estévez and Martínez (1995), Orquera and Piana (1999a)
EASTERN BEAGLE CHANNEL	Alashawaia	Modern	Modern	Modern	8	5.88	12	8.82	116	85.29	0	0	136	Estévez et al. (2001)
	Imiwaia I (K, L, M)	5943 ± 48	6730	6624–6886	3406	7.72	3292	7.46	1814	4.11	35,609	80.7	44,121	Zangrando (2009a), Alunni (2018a, 2018b), Martinoli (2018))

(continued)

Table 3.2 (continued)

Area	Assemblages	Chronology (radiocarbon years BP)	Calibrated years BP (Median probability)	Calibrated years BP: 95.4 (2 sigma)	GUANACOS		PINNIPEDS		BIRDS		FISH		TOTAL NISP	References
					NISP	NISP%	NISP	NISP%	NISP	NISP%	NISP	NISP%		
	Imiwaia I (D)	2987 ± 44 3013 ± 35 3479 ± 150	3109 3147 3707	2959–3247 2999–3253 3514–3831	238	3.28	178	2.46	335	4.62	6497	89.6	7248	Orquera and Piana (2000), Zangrando (2009a), Zangrando et al. (2016)
	Imiwaia I (B)	1495 ± 31	1339	1299–1377	86	1.76	308	6.31	2156	44.19	2329	47.7	4879	Zangrando (2009a), Tivoli (2010a, 2010b), Zangrando et al. (2016)
	Lanashuaia II	1155 ± 40 1385 ± 25	1013 1271	955–1074 1262–1305	0	0	0	0	568	30.74	1280	69.3	1848	Evans et al. (2016)
	Lanashuaia I	1160 ± 70	1027	905–1178	244	3.01	355	4.38	435	5.36	7076	87.3	8110	Piana et al. (2000), Estévez et al. (2001), Mameli and Estévez Escalera (2004), Evans et al. (2016)
	Lanashuaia XXI	825 ± 35	704	663–742	2	1.41	0	0	140	98.59	0	0	142	Álvarez et al. (2013)

(continued)

Table 3.2 (continued)

Area	Assemblages	Chronology (radiocarbon years BP)	Calibrated years BP (Median probability)	Calibrated years BP: 95.4 (2 sigma)	GUANACOS		PINNIPEDS		BIRDS		FISH		TOTAL NISP	References
					NISP	NISP%	NISP	NISP%	NISP	NISP%	NISP	NISP%		
	Heshkaia 34	981 ± 36	851	769–925	1989	78.96	47	1.87	389	15.44	94	3.7	2519	Zangrando et al. (2010), Alunni and Zangrando (2012), Alunni (2018a, 2018b Corbat et al. (2020), Tivoli et al. (2025)
	Heshkaia 35	499 ± 37 532 ± 35 656 ± 35 816 ± 35	509 522 605 702	453–545 496–550 579–654 659–740	3942	57.08	32	0.46	813	11.77	2119	30.7	6906	Zangrando et al. (2014), Alunni (2018a, 2018b), Martinoli (2018), Corbat et al. (2020), Tivoli et al. (2025)
	Heshkaia 28	678 ± 38	605	553–660	405	3.91	163	1.58	329	3.18	9449	91.3	10,346	
	Heskaia 30	263 ± 35	216	142–222	17	1.65	22	2.13	442	42.79	552	53.4	1033	Zangrando et al. (2010), Corbat et al. (2020), Tivoli et al. (2025)
	Heshkaia 39	–	–	–	21	61.76	7	20.59	6	17.65	0	0	34	Zangrando et al. (2010), Alunni and Zangrando (2012)

(continued)

Table 3.2 (continued)

Area	Assemblages	Chronology (radiocarbon years BP)	Calibrated years BP (Median probability)	Calibrated years BP: 95.4 (2 sigma)	GUANACOS		FINNIPEDS		BIRDS		FISH		TOTAL NISP	References
					NISP	NISP%	NISP	NISP%	NISP	NISP%	NISP	NISP%		
SOUTH COAST OF P. MITRE	BVS11 (Early)	5900 ± 60 5085 ± 39 4939 ± 42 4359 ± 43	6681 5807 5639 4905	6496–6799 5703–5907 5578–5744 4327–4993	533	35.99	363	24.51	563	38.01	22	1.49	1481	Vidal (1988), Zangrando et al. (2009), Alunni (2018a, 2018b),Martinoli (2018), Vázquez (2024)
	BVS11 (Middle)	–	–	–	12	26.09	26	56.52	5	10.87	6	13.04	46	
	BVS11 (Late)	550 ± 50 1350 ± 60 1400 ± 34 1590 ± 34	531 1218 1278 1440	439–565 1175–1308 1260–1313 1362–1529	465	51.90	72	8.04	348	38.84	11	1.23	896	
	BVS42	984 ± 36	852	786–926	2636	92.56	39	1.37	166	5.83	7	0.25	2848	Vázquez et al. (2011), Martinoli (2018), Alunni (2018a, 2018b),Vázquez (2024)
	CSM3 (Caleta S. Mauricio)	631 ± 35	603	586–651	0	0	5	71.43	2	28.57	0	0	7	Ceraso et al. (2010)
	BBS S2 (Bahía Buen Suceso)	597 ± 35	552	511–565	0	0	11	100	0	0	0	0	11	Ceraso et al. (2010), Vázquez (2024)

(continued)

Table 3.2 (continued)

Area	Assemblages	Chronology (radiocarbon years BP)	Calibrated years BP (Median probability)	Calibrated years BP: 95.4 (2 sigma)	GUANACOS		PINNIPEDS		BIRDS		FISH		TOTAL NISP	References
					NISP	NISP%	NISP	NISP%	NISP	NISP%	NISP	NISP%		
	BVS11 (Mont. 1)	550 ± 50	531	489–565	340	42.24	239	29.69	220	27.33	6	0.75	805	Vázquez (2024)
	BVS13	370 ± 120	366	238–547	1	0.11	515	55.38	412	44.30	2	0.22	930	Vidal (MS) (in Vázquez et al. (2007)), Vázquez (2024)
	BVS1	335 ± 85	359	250–510	1	0.16	107	16.61	523	81.21	13	2.02	644	Vidal (1985), Vázquez (2024)
	BVS1 (surface)	–	–	–	82	2.95	336	12.09	2358	84.82	4	0.14	2780	Vázquez (2024)
	PSS2 (Punta Sorpresa)	2308 ± 39 369 ± 35	2327 393	2297–2368 311–471	3	0.35	374	43.95	474	55.70	0	0	851	Vázquez et al. (2011), Vázquez (2024)
	PSS1 (Punta Sorpresa)	histórico	histórico	histórico	0	0	77	63.11	16	13.11	29	23.77	122	Ceraso et al. (2010), Vázquez et al. (2011), Vázquez (2024)
	BVS18	histórico	histórico	histórico	7	1.58	8	1.81	90	20.32	338	76.30	443	Vázquez (2024)
	BVS5	–	–	–	98	36.43	60	22.30	111	41.26	0	0	269	
	BVS7	–	–	–	0	0	47	27.33	122	70.93	3	1.74	172	Vidal (1985), Vázquez (2024)

(continued)

Table 3.2 (continued)

Area	Assemblages	Chronology (radiocarbon years BP)	Calibrated years BP (Median probability)	Calibrated years BP: 95.4 (2 sigma)	GUANACOS		PINNIPEDS		BIRDS		FISH		TOTAL NISP	References
					NISP	NISP%	NISP	NISP%	NISP	NISP%	NISP	NISP%		
	BVS9	–	–	–	15	18.29	44	53.66	23	28.05	0	0	82	Vázquez (2024)
	BBS S4 (Bahía Buen Suceso)	–	–	–	0	0	6	28.57	4	19.05	11	52.38	21	Vázquez et al. (2011)
	CH1 (Caleta Herradura)	–	–	–	0	0	15	60.00	10	40.00	0	0	25	
I. DE LOS ESTADOS	Bahía Crossley I	2000 ± 70 2480 ± 60 2730 ± 90	1908 2521 2818	1733–2058 2352–2714 2693–3064	0	0	1130	6.48	16,083	92.3	220	0.13	17,433	Chapman (1987), Horwitz (1990), Horwitz and Weissel (2011), Martinoli (2018)
	Bahía Franklin 3	2070 ± 60	1986	1366–2131	0	0	0	0	328	100	0	0	328	Zangrando et al. (2024a)

In this sector, subsistence strategies display a strong orientation toward guanaco consumption. Moreover, while some assemblages show a predominance of marine resources such as birds, fish, and mollusks, the identified species are mostly associated with intertidal and littoral zones (Zangrando 2010; Zangrando et al. 2010, 2014; Alunni et al. 2020a; Corbat et al. 2020; Tivoli et al. 2025). At the site Heshkaia 30, remains of albatrosses—pelagic birds—have been recorded. However, this may reflect the occasional exploitation of individuals that entered the channel on certain occasions (Tivoli et al. 2025). Additionally, the zooarchaeological remains do not indicate significant consumption of otariids (Martinoli 2018). Meanwhile, cetaceans are present, and their bones show evidence of processing for technological purposes and oil extraction (Martinoli et al. 2022; Zangrando et al. 2024b). In this regard, the available archaeological evidence suggests that marine resource exploitation strategies were likely structured from coastal settings and not necessarily at sea, with a greater reliance on guanacos consumption (Alunni and Zangrando 2012; Zangrando et al. 2014; Alunni 2018a, 2018b; Alunni et al. 2020a).

Along the southern coast of Península Mitre, most archaeological assemblages date to the Late Holocene, with the exception of the Early Component at Bahía Valentín Site 11 (Vidal 1988; Zangrando et al. 2009; Vázquez et al. 2011). In apparent contrast to occupations in the central and eastern sectors of the Beagle Channel, it has been proposed that the proportions of different resources vary according to the situational conditions of each site's location (Vázquez et al. 2011; Vázquez 2024). Indeed, the assemblages exhibit a high degree of variability, suggesting differing organizational circumstances in the use of space and resource procurement (Vidal 1985; Vázquez et al. 2007, 2011; Zangrando et al. 2009; Vázquez 2024). In this regard, it has been suggested that this variability may reflect not only differences in local resource availability, but also a temporal factor linked to the population dynamics of this sector, which may have corresponded to a zone of interaction in which the exploitation ranges of land-based societies from the north and canoe-based societies from the west overlapped (Zangrando et al. 2009; Vázquez 2024).

Finally, in the case of Isla de los Estados—considered an offshore[2] island—(Horwitz and Weissel 2011), the few excavated contexts correspond to the Late Holocene and indicate the exclusive exploitation of marine resources: pinnipeds, cetaceans, fish, and birds, particularly penguins and albatrosses (Horwitz and Weissel 2011; Zangrando et al. 2024a). Among the pinnipeds, subadult males of *Otaria flavescens* predominate (Martinoli 2019). Horwitz (1990) and Horwitz and Weissel (2011) place the technological and faunal data from Bahía Crossley I in a regional context and suggest that Isla de los Estados formed part of a broad maritime circuit that also included other islands such as Nueva, Picton, and Lennox. Furthermore, they propose that this area may have been integrated relatively late into a process of diversification and intensification that occurred in the Fuegian archipelago, particularly in the central Beagle Channel (Horwitz and Weissel 2011; Vázquez 2024).

---

[2] According to Horwitz and Weissel (2011), Isla de los Estados falls within the concept of an *offshore island*, as it is located 24 km away from a larger landmass.

3.5 Zooarchaeological Analysis in Spatial and Chronological Dimensions

As previously stated, the objective of this chapter is to assess the relative contribution of different vertebrate animal resources over time and space along the southern coast of Tierra del Fuego. For this purpose, we employ the Number of Identified Specimens (NISP)—whole bones or fragments identified to taxon. Although this index is subject to various biases (e.g., sampling techniques, taphonomic processes, differential breakage, butchering practices, etc.), it remains one of the most widely used measures in archaeofaunal quantification (Thomas 1969; Grayson 1984; Mengoni Goñalons 1988, 1999; Wheeler and Jones 1989; Lyman 1994, 2003; Reitz and Wing 1999; Gifford-González 2018) and thus enables regional-scale evaluations of vertebrate exploitation. In the context of this study, few works report Minimum Number of Individuals (MNI) per taxon, which makes NISP a more practical tool for incorporating a larger number of sites into the comparative analysis. We acknowledge that differences in the proportions of taxa represented by NISP do not directly reflect caloric contribution or dietary significance; however, we consider it a reasonable proxy for understanding prey selection strategies.

For this chapter, we draw on faunal data from research conducted throughout the history of archaeological investigations in the region (Table 3.2). We analyze four taxonomic categories of vertebrates that are recurrent in the archaeological record: pinnipeds, guanacos, birds, and fish. These categories are present in the vast majority of the studied sites and are commonly reported in terms of their Number of Identified Specimens (NISP), making their inclusion in this comparative study feasible. Taxa that are less frequently represented, such as rodents and canids, are excluded, as their presence may be associated with post-depositional processes (Tivoli and Zangrando 2011). We also do not consider cetacean remains in this study, as current research suggests that many of these may represent by-products related to the production of bone tools. Moreover, the consumption of meat and fat from these animals is difficult to assess, since they may have entered the site without associated skeletal elements (Smith and Kinahan 1984; Borella 2004; Piana 2005; Christensen 2016; Evans et al. 2016).

Regarding the analysis of ichthyofauna, it is important to note that fish remains are particularly sensitive to sampling techniques, as poor visibility during excavation and/or the use of coarse mesh screens often results in the loss of small-sized specimens (Wheeler and Jones 1989; Nagaoka 2005; Zohar and Belmaker 2005). Materials are usually recovered through mesh screens of 5 and 2 mm. Column sampling has been conducted at some of the sites included in this study to allow for a more comprehensive assessment of faunal composition, although this last type of data are not available for all assemblages (Zangrando 2003, 2009a).

This chapter includes 57 faunal assemblages (Table 3.2), comprising a total NISP of 249,900. Of this total, 32.24% corresponds to fish (n = 80,576), 30.07% to pinnipeds (n = 75,146), 27.19% to birds (n = 67,950), and 10.50% to guanacos (n = 26,228).

Following the same spatial segmentation criteria used in previous studies, and considering the percentages of faunal representation, the proportions of different taxonomic groups reveal distinct patterns across the regions of Central Beagle Channel, Eastern Beagle Channel, Península Mitre, and Isla de los Estados (Fig. 3.2). In the central portion of the Beagle Channel, pinnipeds are the most abundant taxon (50.8%), followed by birds (29.1%), fish (11.3%), and guanacos (8.8%). In contrast, in the eastern sector of the Beagle Channel, fish dominate the assemblages (74.4%), followed by guanacos (11.9%), birds (8.6%), and pinnipeds (5.1%). Along the southern coast of Península Mitre, birds comprise the highest proportion of the NISP (43.8%), followed by guanacos (33.7%), pinnipeds (18.9%), and fish (3.6%). On Isla de los Estados, the most frequently represented taxon is birds (92.4%), while pinnipeds account for 6.4% and fish are minimally represented (1.2%).

When analyzing the NISP% values of all assemblages—considering their internal variability and without distinguishing chronological differences—box plot graphs (Fig. 3.3) offer an alternative perspective. In the central portion of the Beagle Channel, pinnipeds exhibit a median of 39.4 with a broad and symmetrical distribution, reaching maximum frequencies close to 90% and 75% of values falling between 1.6 and 69.2%. In the eastern sector of the Beagle Channel, the median for this taxon is much lower (2.3), with a restricted distribution; 75% of the cases fall between 0 and 7.2%, and there is a single outlier at 20%. In contrast, along the southern coast of Península Mitre, pinnipeds have a median of 29.1 and a wide distribution ranging from 1.4 to 100%, with 75% of cases presenting values up to approximately 57.4%. On Isla de los Estados, pinniped remains are found in consistently low percentages. For guanacos in the central portion of the Beagle Channel, the median is 10 and 75% of the assemblages ranging from approximately 0.2 to 23.6%, and a single outlier at

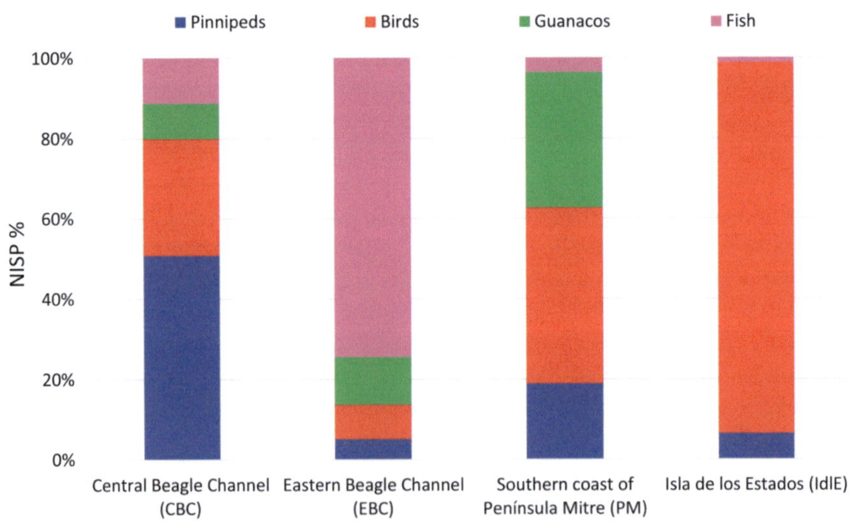

Fig. 3.2 Averaged faunal abundances along the southern coast of Tierra del Fuego

78%. In the eastern sector, the median is 3.6 with an asymmetrical distribution and 75% of the cases falling between 0 and 44.7%. Along the southern coast of Península Mitre, the median is 0.97 with a similarly asymmetrical distribution; 75% of the values range from 0 to 36.1%, and one extreme case exceeds 90%. Guanaco remains are absent from Isla de los Estados. Fish remains show markedly different patterns across regions. In the central portion of the Beagle Channel, the median is 3.6, with an asymmetrical distribution; 75% of values fall between 0 and 17.1%, with outliers reaching between 60% and nearly 80%. In the eastern Beagle Channel, the median rises to 50.6 with a broad distribution; 75% of values fall between 0 and 85.6%. Along the southern coast of Península Mitre, fish are relatively scarce; 75% of the sample records percentages up to 3.1%, though there are several outliers reaching as high as 76.3%. On Isla de los Estados, fish are minimally represented. Finally, regarding birds, the median in the central portion of the Beagle Channel is 16.5, with an asymmetrical distribution and maximum values of 80%; 75% of values range between 2.8 and 39.8%. In the eastern sector, the median is also 16.5, with a skewed distribution, including an outlier at 98.6% and 75% of cases between 3.2 and 43.8%. Along the southern coast of Península Mitre, birds have a median of 33.3, with 75% of the cases falling between 0 and 47.1%. On Isla de los Estados, birds overwhelmingly dominate the assemblages, though this is based on only two cases. In summary, the data reflect variability in resource exploitation across the analyzed coastal sectors. Birds and guanacos exhibit similar trends in the central Beagle Channel, the eastern sector, and Península Mitre, although guanacos reach higher proportions in the latter two. Pinnipeds show greater variability in representation, with a predominance in the central Beagle Channel and high values also observed in Península Mitre. Fish, by contrast, are clearly dominant in the eastern Beagle Channel and decrease markedly in the outer reaches (Península Mitre and Isla de los Estados). Lastly, the pattern observed on Isla de los Estados is consistent with the faunal availability on this island—namely, the absence of guanacos, intensive bird exploitation, and, to a lesser extent, pinnipeds.

In order to further explore relationships among the relative abundances of each taxon, we applied a Principal Component Analysis (PCA). This statistical method aims to simplify the complexity of datasets with many dimensions (in this case, the NISP% of each taxon across different assemblages), while preserving the original information (Carlson 2022). We used the covariance matrix, as all variables are measured on the same scale. The objective of this analysis is to deepen our understanding of the data and identify patterns through integration, considering both spatial provenance and/or chronology. Figure 3.4 displays components 1 and 2, which together account for over 75% of the total variance. The results suggest that faunal abundance patterns cannot be explained considering only the spatial factor. Assemblages with high proportions of fish cluster around the Beagle Channel (both central and eastern sectors), with chronologies mostly from the Late Holocene. Pinnipeds are more abundant in Middle Holocene assemblages from the central Beagle Channel, particularly represented in various phases of the Túnel I Second Component. They are also present in high proportions in the Late Holocene, during later phases of the Península Mitre occupation sequence. Birds are variably present across all coastal

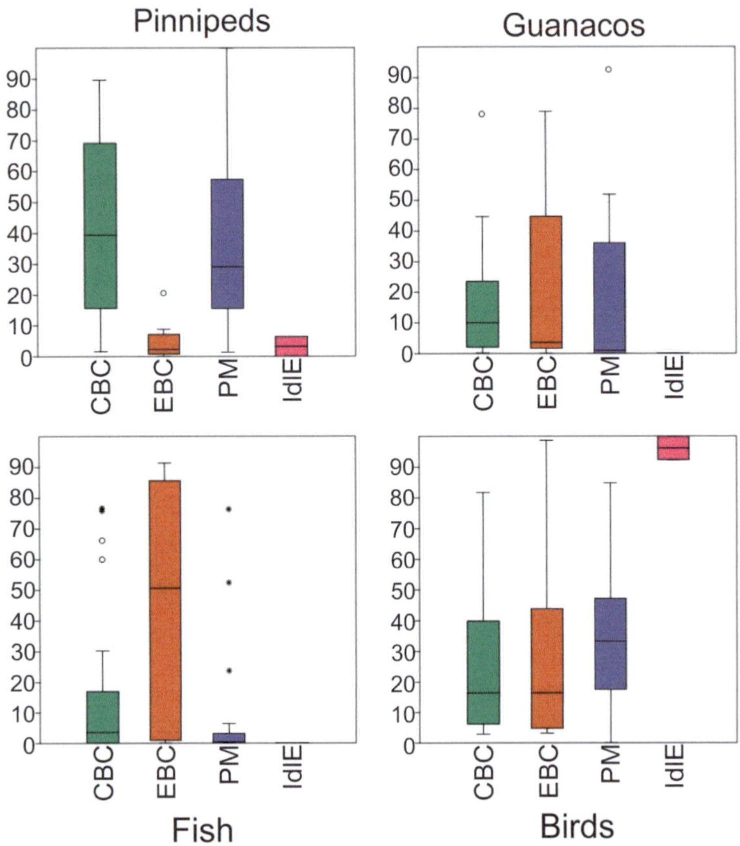

Fig. 3.3 Percentage distribution of faunal remains by taxonomic class across the different archaeological units analyzed. References: (CBC: Central Beagle Channel; EBC: Eastern Beagle Channel; PM: Península Mitre; IdE: Isla de los Estados)

regions but tend to dominate Late Holocene contexts. Guanacos, meanwhile, are distributed throughout the Holocene, without a clear temporal trend, and are found from the central and eastern Beagle Channel to the Península Mitre—though notably absent on Isla de los Estados.

The results of the PCA support some of the spatial trends previously identified, but also indicate that the chronology of the assemblages could be an important factor in accounting for the variability in faunal composition. In this regard, it is important to note that the dispersion of sites along the southern coast of Tierra del Fuego is ubiquitous only from 2000 cal yr BP onward. In earlier periods, the occupation pattern appears more dispersed and focused primarily in the Beagle Channel, from which most of the earliest dates analyzed here originate (ca. 7300–6400 cal yr BP, mainly represented by various phases of Túnel I and layers K, L, and M of Imiwaia I), although Península Mitre also records a site (BVS11 Early Component) dated to

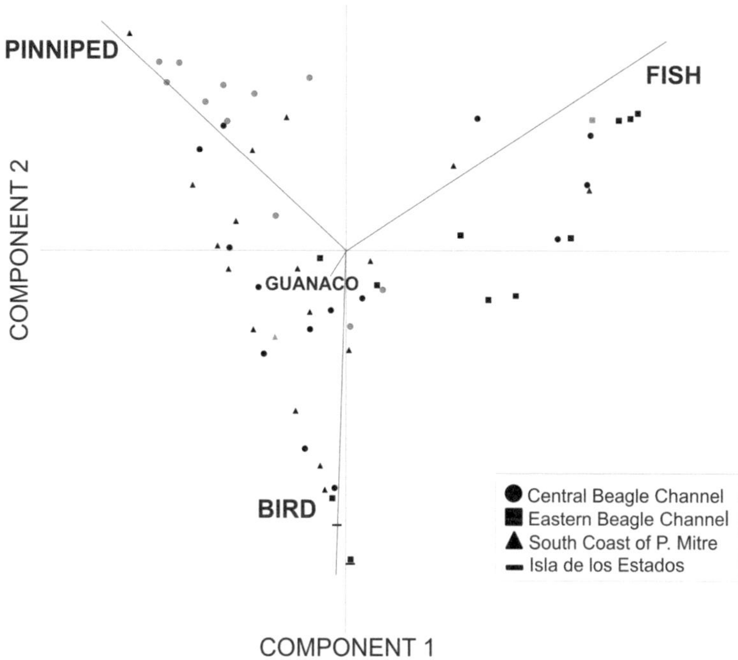

Fig. 3.4 Principal Component Analysis (PCA). References: orange: Middle Holocene assemblages, black: Late Holocene assemblages

6600 cal yr BP. Dates from the interval between 5700 and 3000 cal yr BP predominantly correspond to the Beagle Channel and a few dates from BVS11 (Península Mitre). Afterward, during the period 3000–2000 cal yr BP, there are records of a few occupations on Península Mitre (Punta Sorpresa 2) and in the Beagle Channel. Additionally, the first human evidence from Isla de los Estados was found, with a date of 2700 cal yr BP (Bahía Crossley I). After 2000 cal yr BP, there is an increase in the number of sites along the entire southern coast of Tierra del Fuego.

We then evaluated the representation of fauna across a temporal continuum, as shown in Fig. 3.5. Middle Holocene sites exhibit a predominance of pinniped remains over other taxonomic groups (which correspond to phases of the Second Component of Túnel I), while birds are proportionally more significant in Late Holocene assemblages. By this time, fish also show a gradual increase in abundance, while guanacos maintain relatively stable proportions throughout the occupation sequence. However, in the Late Holocene, several cases with high relative frequencies of guanacos are recorded. Therefore, greater variability in faunal proportions is inferred for the Late Holocene, which could reflect changes in subsistence patterns due to demographic growth at a regional scale (Zangrando 2009a, 2009b; Tivoli and Zangrando 2011; Vázquez and Borrero 2021). The increase in archaeological evidence after 2000 cal yr BP and the proliferation of sites in the eastern Beagle Channel and Península Mitre (Vázquez and Borrero 2021; Vázquez 2024) may reflect this process.

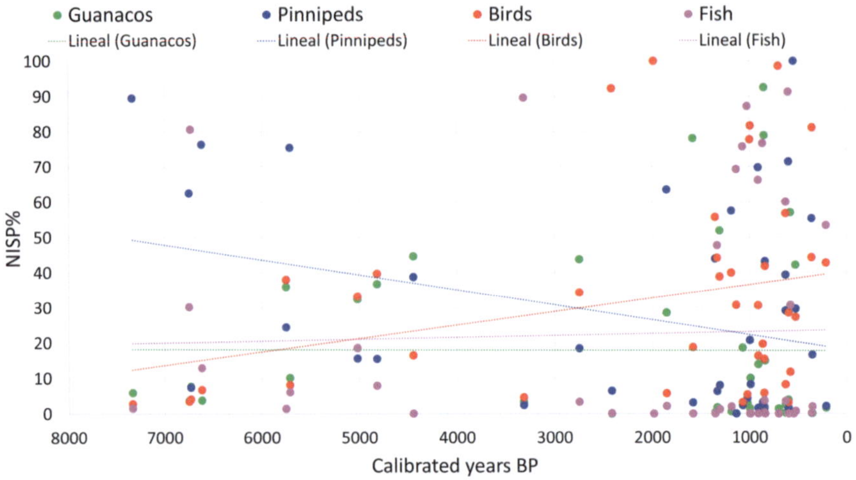

Fig. 3.5 Faunal abundances from the southern coast of Tierra del Fuego represented according to the calibrated age of the assemblages

We therefore aimed to assess the degree of resource dominance over time, using the equitability or evenness index (V') (Reitz and Wing 1999). The resulting values range from 0 to 1: those close to zero indicate assemblages in which one of the taxonomic categories dominates, while values close to 1 indicate assemblages in which there is an even distribution of the different taxonomic categories (Reitz and Wing 1999; Tivoli and Zangrando 2011). As shown in Fig. 3.6, the index reveals some temporal and spatial variations. Most assemblages dated prior to 5700 cal yr BP record intermediate values (between 0.3 and 0.65). These include assemblages corresponding to the Second Component phases of Túnel I, with a large amount of pinnipeds, and the assemblage corresponding to the Middle Holocene levels of Imiwaia I, with a large quantity of fish remains. Within this same temporal block, but differing from the aforementioned, is the Early Component of BVS11 (Península Mitre), where birds, guanacos, and pinnipeds are represented in similar proportions (V' = 0.82), indicating a more taxonomically uniform assemblage. Between 5000 and 4000 cal yr BP, the indices show values closer to 1, reflecting a more uniform distribution in the assemblages (Fig. 3.5). From 3000 cal yr BP onward, the assemblages exhibit a broader range in the index values (Fig. 3.6); indeed, both assemblages dominated by a particular taxon and others with more uniform distributions are observed. This suggests, on one hand, a less uniform resource selection—more focused on high-return resources—in the older assemblages, and on the other hand, greater variability in the composition of faunal assemblages from more recent chronologies (Figs. 3.6 and 3.7). An increase in the archaeological signal can also be observed in the Late Holocene in all the sectors analyzed, as well as the beginning of the archaeological signal on Isla de los Estados (Fig. 3.7).

From the Late Holocene onwards, several aspects require consideration with regard to the different spatial contexts from which the assemblages originate, and with

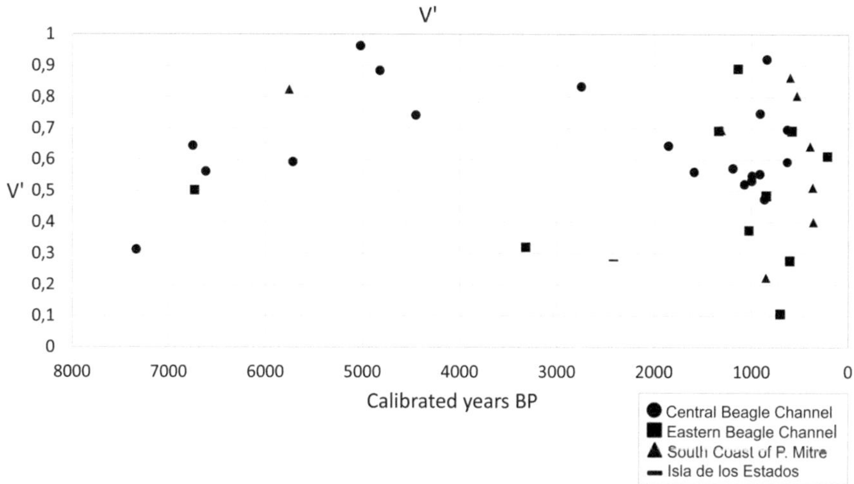

Fig. 3.6 Equitability (or evenness) index calculated for the assemblages from the southern coast of Tierra del Fuego, represented as a function of their calibrated ages

respect to the different vertebrate taxa analyzed: (1) the assemblages from Isla de los Estados are relatively low in taxonomic diversity, with birds overwhelmingly predominant (Figs. 3.6 and 3.7); (2) both in the Beagle Channel and Península Mitre, the zooarchaeological assemblages display a wide range of variability, including assemblages with a uniform distribution of taxonomic proportions and others in which one or another taxon predominates without a consistent pattern (exceeding 50% of total NISP%), thus reflecting a high diversity of subsistence strategies (Figs. 3.6 and 3.7); (3) most Late Holocene cases with lower evenness index values (between 0 and $\leq$ 0.50) are assemblages dominated by fish or birds (Fig. 3.6), with the exception of BVS42 (Península Mitre), which is dominated by guanacos; (4) in the sites from Península Mitre and Isla de los Estados, fish remains are very scarce (Fig. 3.7), a pattern that may be related to sampling issues.

3.6 Faunal Assessment in a Broad Regional Context

The results derived from the integration of assemblages originating from the various areas that constitute the southern coast of Tierra del Fuego indicate a certain degree of variability at both spatial and temporal scales. Therefore, in the following section, we will discuss both dimensions of vertebrate resource exploitation.

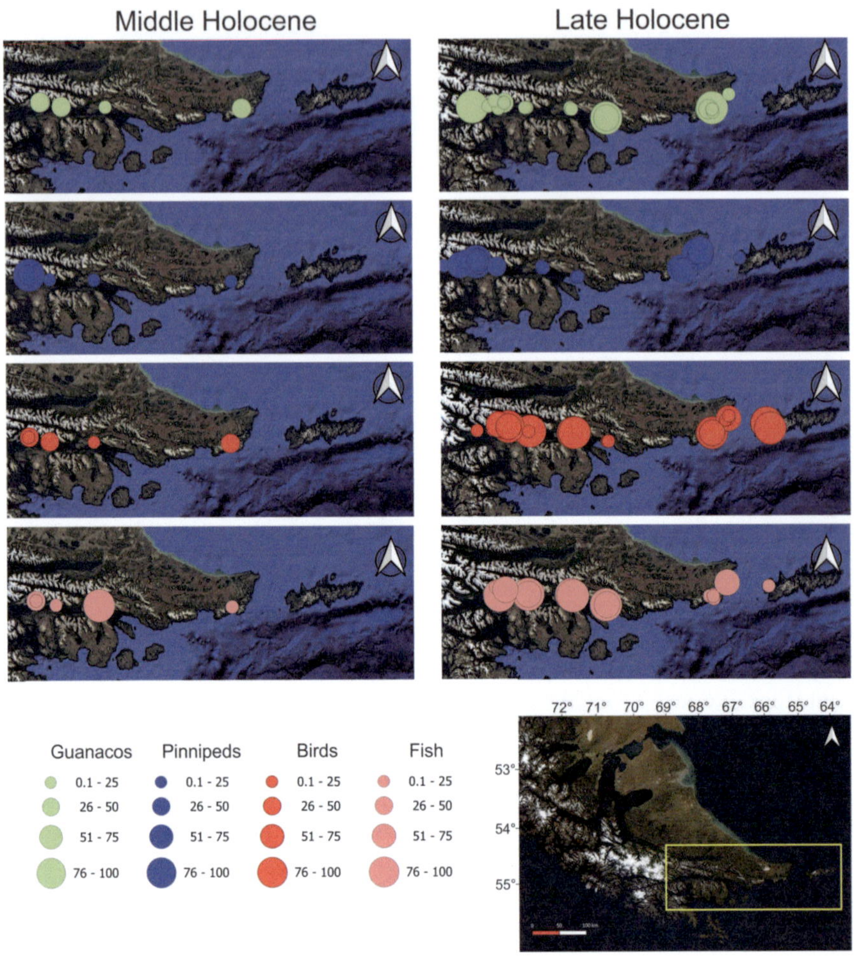

Fig. 3.7 Map of faunal abundances (NISP %) on the southern coast of Tierra del Fuego

3.6.1 Faunal Exploitation in a Spatial Dimension

In general terms, the zooarchaeological analysis of the southern coastal region reveals a strong reliance on marine resource consumption. This pattern is consistent with other lines of evidence, such as stable isotope studies, which have shown that $\delta^{13}C$ and $\delta^{15}N$ values in human remains from the Beagle Channel and Península Mitre indicate predominantly marine-based diets, albeit with some contribution from terrestrial animal fats (Kochi et al. 2018; Alunni et al. 2021; Zangrando et al. 2022).

However, some differences have been observed regarding the relative abundance of various taxa along the southern coast of Tierra del Fuego. One possible explanation lies in the differential availability of resources, particularly pinnipeds and guanacos.

In this regard, two aspects are worth highlighting: (1) the west–east decrease in mountain elevation may have facilitated guanaco and human movement between the coast and the interior in the eastern sectors of the Beagle Channel and Península Mitre; and (2) the outer zones of the Fuegian archipelago, such as Península Mitre and Isla de los Estados, exhibit high marine productivity and may have offered a different diversity and abundance of resources compared to areas within the Beagle Channel, particularly in relation to marine mammal and seabird colonies (Acha et al. 2004; Schiavini and Raya Rey 2001). Therefore, the following section explores variations in resource exploitation strategies, since although all of these taxa are present across the Fuegian archipelago (with the exception of guanacos on Isla de los Estados), differences in their productivity and distribution may have prompted distinct adaptive responses among human groups (Erlandson 2012).

In the case of pinniped exploitation, assemblages from the Beagle Channel exhibit a representation of nearly all age and sex categories of *Arctocephalus australis*, with a slight predominance of adult and subadult males, although females and individuals younger than one year are also present (see following section), along with a low frequency of *Otaria flavescens* remains. Furthermore, most anatomical units are represented in roughly similar proportions, suggesting that complete carcasses were transported to the archaeological sites (Martinoli 2017, 2018, 2019). Both mortality profiles and anatomical distributions have been interpreted as the result of aquatic hunting, facilitated by the use of watercraft and harpoons with detachable points (Schiavini 1990, 1993; Orquera and Piana 1999a). However, as will be discussed below, from the Late Holocene onward, patterns of capture, processing, and consumption became more variable (Martinoli 2018). In the specific case of Moat Bay, in the eastern Beagle Channel, remains of *O. flavescens* and *A. australis* are found in similar proportions, although there is a higher representation of male individuals of the former species—animals which can exceed 300 kg in weight. The anatomical profiles suggest differential transport of body parts (Martinoli 2018). It has been proposed that these resources were likely hunted on land, considering the lack of evidence for specialized maritime hunting technologies in the studied assemblages from this sector (Zangrando et al. 2014; Martinoli 2018). Meanwhile, studies conducted on assemblages from Bahía Valentín, in Península Mitre, indicate heterogeneous patterns of pinniped exploitation depending on site location, which may reflect mixed strategies involving both aquatic and terrestrial hunting (Martinoli 2018; Martinoli and Zangrando 2024). At site BVS11, both pinniped species are present in similar proportions, with all age and sex categories represented. In contrast, evidence from BVS42 suggests the transport of relatively incomplete carcasses into the interior of Bahía Valentín (Martinoli 2018). Finally, in Isla de los Estados assemblages, *O. flavescens* predominates, particularly subadult males (approximately 200 kg), which likely implies both aquatic and terrestrial hunting at haul-out and/or breeding colonies (Martinoli 2018, 2019). In summary, the mortality profiles, transport strategies, and differential prey sizes point to a trend toward more systematic exploitation of animals on land (resting and/or breeding colonies) from Moat Bay eastward (Martinoli 2017, 2018, 2019; Vázquez 2024).

Regarding guanaco hunting, previous studies indicate a degree of variability in transport and processing strategies across different areas (Vázquez 2015; Alunni 2018a, 2018b, 2021). On the one hand, differences have been proposed in the foraging ranges for guanaco procurement: areas near base camps in assemblages such as Layer D of Túnel I (Vázquez 2015), and broader foraging territories in sites from Moat Bay (Alunni et al. 2020a). On the other hand, with respect to bone processing and final consumption practices, some assemblages from the central and eastern Beagle Channel—specifically the Fourth Component of Túnel I and the Bahía Moat sites (Heshkaia 34 and 35)—dated to the Late Holocene, suggest energetically costly extraction activities. Similar patterns are also seen in the lower component (Layers K, L, and M) of Imiwaia I, dated to the Middle Holocene. These practices are associated with the processing of bone grease contained in the trabecular tissue of the hunted guanacos (Alunni et al. 2020a, 2021; Alunni 2021; Alunni and Zangrando 2024). Such activities likely involved the transport and exploitation of anatomical parts with extremely marginal economic returns and high processing demands—particularly for non-ceramic societies. In contrast, in the southern coast of Península Mitre, guanacos are found in high proportions in sites BVS11 and BVS42, especially in the layers corresponding to the Late Holocene (ca. 1900–750 cal yr BP), where lithic technologies are consistent with the capture of these camelids (e.g., stone projectile points), and where bone technology is relatively scarce (Vázquez 2024). Conversely, guanaco remains have not been recorded in assemblages near the coast of the Le Maire Strait. For Península Mitre, it has been proposed that guanacos were exploited from areas close to residential bases (Alunni 2018a, 2021; Alunni et al. 2020a). In this context, differential transport of anatomical parts may have been linked to the selection of elements suitable for marrow extraction or feasible to be used in the production of bone technology (Scheinsohn 1997, 2010; Christensen 2016; Alunni et al. 2020b). Notably, no evidence of bone grease extraction has been documented in any assemblage from Península Mitre (Alunni et al. 2021). The observed differences may be related to varying levels of access to these prey species by hunter-gatherer groups. In the southern coast of Península Mitre, guanacos may have provided a more consistent resource both seasonally and geographically, contrasting with what appears to have occurred in the central Beagle Channel (Orquera and Piana 1999b; Alunni et al. 2020b, 2021; Vázquez 2024). This greater availability of guanacos—combined with access to otariids colonies on land, as discussed above—may have allowed for the continuous procurement of both fats and proteins throughout the annual cycle.

Regarding fish, studies conducted in the Beagle Channel have shown the consumption of various species, including both coastal (particularly notothenioids) and pelagic species (Juan-Muns i Plans 1992, 1996; Zangrando 2003, 2009a, 2009b). However, in the case of the Moat Bay sites, the species recorded are all coastal or intertidal, primarily from the notothenioid group (Corbat et al. 2020). Likewise, both the southern coast of Península Mitre and Isla de los Estados present fish remains in very low percentages (Chapman 1987; Horwitz 1990; Horwitz and Weissel 2011; Vázquez 2024).

Birds play a prominent role in the zooarchaeological assemblages from Isla de los Estados (Horwitz and Weissel 2011; Zangrando et al. 2024a, b), while showing variability in their representation proportions in the Beagle Channel and Península Mitre (Mameli 2000, 2003; Mameli and Estévez Escalera 2004; Tivoli 2010a, 2010b, 2012, 2014; Vázquez et al. 2011; Zangrando et al. 2010, 2014; Tivoli et al. 2025). In the assemblages of Isla de los Estados, penguins are particularly abundant, followed by significant proportions of albatrosses, cormorants, and a smaller number of geese (cauquenes) (Horwitz and Weissel 2011; Zangrando et al. 2024a). On the southern coast of Península Mitre, identifications indicate the predominant presence of cormorants and penguins, including a significant proportion of *Aptenodytes* sp. remains. Seagulls and various procellarids appear in lower frequencies, while other categories such as anatids and falconids were identified in minimal proportions (Vázquez et al. 2011; Vázquez 2024). Finally, in the Beagle Channel archaeological record, penguins, procellarids (both small and large), and especially cormorants, are the most abundant species, though anatids, seagulls, some raptors, and other species with lower representation are also typically present (Estévez Escalera et al. 2001; Mameli and Estévez Escalera 2004; Tivoli 2010a, 2010b, 2012, 2014; Tivoli et al. 2025).

When examining the faunal information at the proposed regional scale and in conjunction with the inferred exploitation modes to date, several observations can be made. The results presented here indicate that assemblages in which pinnipeds predominate are primarily located in Península Mitre and the central portion of the Beagle Channel, whereas in the eastern sector of the Channel, they are represented in smaller proportions. Morover, both in the central Beagle Channel and in Península Mitre, procurement strategies suggest that the hunting capacity allowed for the exploitation of these animals both in the water and on land. Specifically, on the southern coast of Península Mitre, the land-based option was likely more common due to the greater availability of otariid aggregation areas. In the case of Bahía Moat, provisioning strategies indicate that animals were mostly exploited in land-based spaces, although the representation of pinnipeds in this sector is low (Martinoli and Zangrando 2024). In the case of Isla de los Estados, exploitation also appears to have occurred both on land and in the water, and although pinnipeds are not abundant in terms of NISP%, the large size of the selected prey indicates they were likely important resources in the diet. Regarding guanacos, they are present in sites from both the Beagle Channel and Península Mitre, with some sites showing their predominance over other taxa. In the particular case of the eastern Beagle Channel, the percentage of these camelids is comparatively higher, especially in Moat Bay, where they were intensively processed for the extraction of bone grease (Alunni 2021; Alunni et al. 2021). In relation to fish, they present relatively high values across the Beagle Channel, particularly in the eastern sector (represented mainly by the Imiwaia I assemblages). However, there are differences regarding the selected species, as while in the central Beagle Channel both coastal and pelagic taxa were consumed, in the eastern Beagle Channel, differences are noted depending on the site location: at Imiwaia I, both coastal and pelagic fish are recorded, but at Moat Bay, only coastal species were captured. On the southern coast of Península Mitre and

Isla de los Estados, the percentage of fish is generally very low; however, it should be considered that this could be linked to excavation strategies. Finally, birds are present in nearly all assemblages, with a predominance of cormorants and penguins, although in some cases, high proportions of procellarids are recorded. Other taxa generally appear in low proportions. The predominance of birds is particularly notable in the assemblages from Isla de los Estados.

3.6.2 Temporal Dimension and the Perspective of Change

This section examines chronological variation in the archaeological assemblages, as previous analyses suggest that patterns of resource exploitation may be partially linked to changes occurring over time.

According to the results, most archaeological assemblages dated between 7300 and 5700 cal yr BP tend to be dominated by pinnipeds, with the exception of Middle Holocene contexts from BVS11 and Imiwaia I. However, this assertion must be approached with caution, given that most of the early-dated assemblages correspond to different phases of the Second Component of Túnel I. This situation could be overestimating the importance of pinnipeds during these phases of the occupational sequence, as sites such as Imiwaia I (Layers K, L, and M) and the early component of BVS11 present pinniped NISP percentages (7.5% at Imiwaia I and 23.6% at BVS11) comparable to those of guanacos (7.7% at Imiwaia I and 35.6% at BVS11).

Moreover, although Fig. 3.5 shows a declining trend in pinniped consumption toward the Late Holocene, this resource continues to be present in the majority of sites, and in some assemblages it even predominates (e.g., Ajej I, Layer C of Mischiuen I, BVS13, PSS1, among others).

A substantial temporal difference regarding pinniped exploitation lies in the modes through which these animals were targeted (Martinoli 2017, 2018; Martinoli and Zangrando 2024). During the Middle Holocene at Túnel I and Imiwaia I, the majority of identified individuals were adult and subadult male *Arctocephalus australis* that died during the winter, leading to the proposal that they were hunted at sea outside of the reproductive season (Schiavini 1990, 1993). However, these profiles begin to shift from the end of the Middle Holocene onward (Martinoli 2018). Indeed, previous analyses have proposed that the exploitation strategies of this high-ranked resource during the Middle Holocene occurred within a context involving low-density hunter-gatherer populations with strong social and reproductive ties to other groups in the Strait of Magellan/Otway Sound region (Fiore 2006; Álvarez 2004, 2009a, b; Zangrando 2009a:105; San Román 2013). These inter-regional networks of mobility and communication are thought to have extended as far as the southern coast of Península Mitre, as evidenced by the composition of the Early Component of BVS11, where both guanacos and pinnipeds were significant prey (Zangrando et al. 2009; Vázquez 2024). Nevertheless, the scarcity of Middle Holocene sites in Moat Bay and Península Mitre suggests that occupational intensity in these areas may have been significantly lower during this phase of regional settlement. However, this pattern remains tentative, and future surveys in these regions may provide new insights into human occupations during this period.

Assemblages dated between approximately 5700 and 3000 cal yr BP show an increase in evenness values (Fig. 3.6), suggesting greater variability in resource exploitation strategies. This trend reflects the increasing frequency of certain prey types that had been previously exploited. In addition, changes in pinniped procurement strategies are observed: there is a marked increase in the capture of adult females and individuals under one year of age of *Arctocephalus australis*, that is, smaller prey more commonly available at haul-out and/or breeding colonies (Martinoli 2018; Martinoli and Zangrando 2024). Thus, not only were some taxa more intensively exploited, but capture strategies also varied, implying the incorporation of new procurement zones (Tivoli and Zangrando 2011; Martinoli 2018). These changes can be conceptualized within the framework of a dietary broadening process. According to Zangrando (2009a, 2009b), such developments began around 5500 cal yr BP, within a phase of regionalization and spatial circumscription of groups already established in the area. These processes may be linked to increasing demographic growth, which could have led to greater stress on high-ranked resources such as pinnipeds (Nye et al. 2020; see Chap. 2). A potential response to these conditions may have involved shifts in hunting strategies and technological innovations aimed at enhancing efficiency in resource acquisition (Winterhalder et al. 1988; Broughton 1994, 1997; Broughton and O'Connell 1999; Lyman 2003; Scheinsohn 2010, among others). Such dynamics appear to be reflected in the greater diversity of age/sex categories and taxa exploited during this period.

Around 3000 cal yr BP—and particularly after 2000 cal yr BP—new variations are observed in faunal indices: evenness values exhibit a wide range of taxonomic representations, from assemblages dominated by a single taxon to others with greater diversity. Simultaneously, birds gain prominence across all coastal areas, in contrast to their relative scarcity in Middle Holocene assemblages. Once again, changes in faunal abundance are accompanied by shifts in resource procurement strategies. Among fish, analyses of assemblages from the Beagle Channel have identified a significant increase in the capture of deep-water species, as well as a rise in the exploitation of smaller Nototheniidae (*P. magellanica* and *Patagonotothen* sp.) (Zangrando 2009a, 2009b; Tivoli and Zangrando 2011). However, it is noteworthy that in Bahía Moat—in the eastern Beagle Channel—no deep-water species were identified, although Nototheniidae were clearly exploited (Corbat et al. 2020). This suggests: (1) an expansion of foraging ranges (Zangrando 2009a), and (2) the use of fishing technologies requiring high labor investment, within a socio-economic context that may have fostered innovations in capture strategies, such as the construction of fish traps (Vázquez and Zangrando 2017). Avifaunal assemblages from the Beagle Channel also show changes in species representation, with a greater inclusion of pelagic birds such as albatrosses, petrels, and shearwaters, along with evidence of sites likely associated with specialized cormorant exploitation (Piana et al. 2007a; Piana and Vázquez 2008; Tivoli and Zangrando 2011; Tivoli 2012, 2014). Examples such as Túnel II (Piana and Canale 1993–94), Lanashuaia XXI (Álvarez et al. 2013), and Shamakush VIII (Piana and Vázquez 2008), where there is an overwhelming accumulation of remains of cormorants, may represent such strategies. Similarly, in Península Mitre, an accumulation of large Procellariiformes (possibly Black-browed Albatrosses) was documented at PSS2, alongside other cases that suggest strategies for obtaining multiple prey (Vázquez et al. 2007; Vázquez 2024). Regarding pinniped exploitation, mortality profiles in the central Beagle Channel indicate a greater diversity than in earlier periods, with some assemblages dominated by adult females and pups, as well as year-round hunting (Piana et al. 2008; Martinoli and Vázquez 2017; Martinoli et al. 2021). For this period, an intensification process has been proposed for the Beagle Channel (Zangrando 2009a, 2009b), situated within the framework of effective spatial occupation (sensu Borrero 1989–90). This process was characterized by increased labor investment in resource acquisition, changes in their relative abundances, and the incorporation of new procurement zones (Zangrando 2009a, 2009b; Tivoli and Zangrando 2011). In the southeastern coast of Península Mitre, the archaeological signal is relatively weak and discontinuous between 3000 and 1000 cal yr BP. Only around 750 cal yr BP does the evidence become more intense and geographically consistent. Assemblages from this later phase display considerable variation in faunal representation, with a general orientation toward increased use of low return rates resources, particularly birds. Such findings have been understood as indicative of a delayed process of effective space occupation compared to the central sector of the Beagle Channel (Vázquez 2024).

Among the possible causes of these diachronic changes, it has been proposed that increasing population density may have resulted in renewed pressure on exploited resources, leading to the more frequent incorporation of lower-ranked species and promoting the expansion into previously under-occupied areas (Legoupil and Fontugne 1997; Scheinsohn 1997, 2010; Zangrando 2009b; Tivoli and Zangrando 2011).

3.7 Concluding Remarks

The analyses conducted in this study provide a regional perspective on the patterns of change and continuity in subsistence strategies throughout the Holocene. This perspective is grounded in an integrated examination of zooarchaeological assemblages recovered from the southern coast of Tierra del Fuego, dating from approximately 7300 cal yr BP onward. Adopting this broad-scale approach enables us to propose a connection between regional settlement dynamics, the use of coastal environment, and specific exploitation strategies implemented in each sector, as shaped by the availability and distribution of targeted faunal taxa (Erlandson 2012).

Between ca. 7300 and 5700 cal yr BP, pinnipeds emerged as the predominant resource, likely due to their high return rates (Zangrando 2009a). Hunting activities during this period focused particularly on adult and subadult males of *Arctocephalus australis*. The zooarchaeological assemblages dated to this time span originate almost exclusively on the Beagle Channel, with the exception of the Early Component of

BVS11 in Península Mitre, where a significant exploitation of high-return prey—otariids and guanacos—is also evident. In this context, the Beagle Channel (along with the Strait of Magellan/Otway region) has been proposed as a "population core" from which human groups gradually expanded toward the outer arc of the archipelago, which was populated at a later stage (Legoupil and Fontugne 1997; Eisdhaug 2024). Between 5700 and 4000 cal yr BP, assemblages from the Beagle Channel exhibit increased variability in consumed resources, along with evidence for emerging strategies of pinniped exploitation. After 3000 cal yr BP, the archaeological signal begins to expand along the entire southern coast, with more redundant patterns of spatial occupation, particularly from 2000 cal yr BP onward (Fiore 2006; Álvarez 2003; Zangrando et al. 2009; Zangrando et al. 2010; Horwitz and Weissel 2011; Vázquez 2024). The zooarchaeological record of these later assemblages reveal several broadly shared patterns: (1) a high degree of faunal variability, with sites dominated by a single taxon and others characterized by more evenly distributed resource use; and (2) a general increase in the frequency of birds, which had not been prominent in earlier coastal contexts. Nevertheless, some spatial distinctions persist: the central sector of the Beagle Channel displays a higher proportion of pinniped remains, although Late Holocene assemblages in this area show greater variability than those from earlier periods. In contrast, the eastern sector presents higher proportions of fish and guanacos and relatively low frequencies of pinnipeds. Along the southern coast of Península Mitre, relative abundances vary, with assemblages alternately dominated by different taxa. Finally, Isla de los Estados appears to exhibit a degree of specialization in bird exploitation, with pinnipeds incorporated to a lesser extent and in lower proportions.

The southern coast of Tierra del Fuego is a region characterized by a relatively homogeneous resource base, implying that movement between sectors did not entail significant ecological resistance (sensu Erlandson 2012). Indeed, the southern coast of Península Mitre was already part of the mobility circuit as early as the Middle Holocene, albeit likely within a context of low demographic density (Zangrando et al. 2009; Vázquez 2024). The availability of watercraft likely enabled the development of broad mobility ranges (Ames 2002), thereby incorporating areas that might otherwise be considered "undesirable" (sensu Scheinsohn 1997, 2010), such as Isla de los Estados and other islands of the Fuegian archipelago. These spaces appear to have become increasingly integrated into mobility and communication networks as pressure on both resources and territory intensified (Horwitz and Weissel 2011; Martinoli 2019; Zangrando et al. 2024a, b). By the Late Holocene, new variations in subsistence strategies become evident: (1) more systematic exploitation of reproductive and/or haul-out pinniped colonies; (2) the consumption of bird and fish species from pelagic zones; (3) intensive exploitation of small-sized coastal fish; and (4) a certain degree of specialization in bird procurement. These temporal trends observed along the southern coast of Tierra del Fuego likely reflect broader processes of population expansion and settlement, wherein the occupation of different coastal spaces occurred at distinct tempos. Considering the southern Fuegian coast as an integrated system has allowed for the identification and description of temporal trends in faunal abundances and exploitation strategies, within the broader framework of human use of coastal environments tied to different stages of the Fuegian littoral occupation process. Resource exploitation strategies during the earliest recorded phase (ca. 8600–7300 cal yr BP) remains poorly understood, as the fragmentary nature of the zooarchaeological record currently precludes detailed assessment of this time span.

References

Acha E, Mianzan H, Guerrero R, Favero M, Bava J (2004) Marine fronts at the continental shelves of austral South America: physical and ecological processes. J Mar Sys 44(1):83–105. https://doi.org/10.1016/j.jmarsys.2003.09.005

Alunni DV (2018a) Estrategias de aprovisionamiento de guanacos y uso del bosque Magallánico por cazadores–recolectores costeros. PhD Dissertation, Universidad de Buenos Aires

Alunni DV (2018b) Estrategias de aprovisionamiento de guanacos y uso del bosque Magallánico por cazadores–recolectores costeros. Rev Mus Antrop 11(2):7–22. https://doi.org/10.31048/1852.4826.v11.n2.18669

Alunni D (2021) Ungulates butchering and transport by hunter–gatherers with maritime economic orientation: the case of the south coast of Tierra del Fuego (Argentina). J Anthropol Archaeol 64:101336. https://doi.org/10.1016/j.jaa.2021.101336

Alunni DV, Zangrando AF (2012) Primeros datos sobre el transporte, procesamiento y consumo de guanacos en la localidad arqueológica Heshkaia (sudeste de Tierra del Fuego, Argentina). Magallania 40(1):319–331. https://doi.org/10.4067/S0718-22442012000100019

Alunni DV, Zangrando AF (2024) Axial skeletal bone grease rendering in guanacos: implications for the zooarchaeological record of Patagonia and Tierra del Fuego (Argentina). J Archaeol Sci Rep 54:104462. https://doi.org/10.1016/j.jasrep.2024.104462

Alunni DV, Tivoli AM, Zangrando AF (2020a) Movilidad y explotación de recursos terrestres por cazadores–recolectores costeros de Bahía Moat, costa sureste de Tierra del Fuego, Argentina. Lat Am Antiq 31(3):576–594. https://doi.org/10.1017/laq.2020.27

Alunni DV, Vázquez M, Zangrando AF (2020b) Explotación de guanacos por cazadores–recolectores durante el Holoceno medio (Argentina). Arqueología 26(2):157–180. https://doi.org/10.34096/arqueologia.t26.n2.5826

Alunni DV, Kochi S, Tessone A, Zangrando AF (2021) Consumo de grasas entre cazadores–recolectores de la costa sur de Tierra del Fuego durante el Holoceno tardío: una evaluación desde la zooarqueología y los isótopos estables. Comechingonia 25(3):195–228. https://doi.org/10.37603/2250.7728.v25.n3.32419

Álvarez MR (2003) Organización Tecnológica en el Canal Beagle. El Caso de Túnel I (Tierra del Fuego, Argentina). PhD Dissertation, Universidad de Buenos Aires

Álvarez M (2004) Estrategias tecnológicas en los grupos canoeros tempranos del área Fuego-Patagónica. Magallania 32:191–208. https://www.magallania.cl/index.php/magallania/article/view/1208

Álvarez MR (2005) Organización tecnológica en la costa norte del canal Beagle: estrategias de uso de materiales líticos de los cazadores litorales del extremo sur de Patagonia. Rel Soc Argent Antropol XXX:33–57

Álvarez MR (2009a) Tendencias y cambios en las prácticas tecnológicas de los grupos cazadores-recolectores del extremo sur sudamericano. In: Salemme M, Santiago F, Álvarez M, Piana E, Vázquez M, Mansur E (eds) Arqueología de la Patagonia: una mirada desde el último confín, tomo II. Utopías, Ushuaia, pp 19–33

Álvarez MR (2009b) Diversidad tecnológica en el extremo sur de Patagonia: tendencias y continuidades en el diseño y uso de materiales líticos. In: Barberena R, Borrazzo K, Borrero LA (eds) Perspectivas Actuales en Arqueología Argentina, CONICET–IMHICIHU, Buenos Aires, pp 241–267

Álvarez M, Fiore D, Tivoli AM, Salvatelli L, Saletta MJ, Briz I (2013) Variabilidad de actividades humanas en momentos recientes de la ocupación del canal Beagle (Tierra del Fuego): el caso de Lanashuaia XXI. In: Zangrando AF, Barberena R, Gil A, Neme G, Giardina M, Luna L, Otaola C, Paulides S, Salgán L, Tivoli AM (eds) Tendencias Teórico Metodológicas y Casos de Estudio en la Arqueología de Patagonia. Sociedad Argentina de Antropología, INAPL, Museo de Historia Natural de San Rafael, Buenos Aires, pp 559–568

Ames K (2002) Going by Boat. The Forager–Collector Continuum at Sea. In: Fitzhugh B, Habu J (eds) Beyond foraging and collecting. Evolutionary change in Hunter–Gatherer settlement system. Kluwer Academic/Plenum Publishers, New York, pp 19–52

Bailey G (2008) Time perspectivism: origins and consequences. In: Holdaway S, Wandsnider L (eds) Time in archaeology: time perspectivism revisited. The University of Utah Press, Salt Lake City, pp 13–30

Borrero LA (1989–90) Evolución cultural divergente en la Patagonia austral. Anal Inst Pat 19:133–140

Borella F (2004) Tafonomía regional y estudios arqueofaunísticos de cetáceos en Tierra del Fuego y Patagonia meridional. BAR, Oxford

Borromei A, Ponce JF, Coronato A, Candel M, Olivera D, Okuda M (2014) Reconstrucción de la vegetación posglacial y su relación con el ascenso relativo del nivel del mar en el extremo este del canal Beagle, Tierra del Fuego Argentina. Andean Geology 41(2):362–379. https://doi.org/10.5027/andgeoV41n2-a05

Bowen W, Beck C, Austin D (2009) Pinniped ecology. In: Perrin W, Würsig B, Thewissen J (eds) Encyclopedia of marine mammals, 2nd edn. Academic Press, New York, pp 852–861

Broughton J (1994) Late Holocene resource intensification in the Sacramento Valley, California: the vertebrate evidence. J Archaeol Sci 21:501–514. https://doi.org/10.1006/jasc.1994.1050

Broughton J (1997) Widening diet breadth, declining foraging efficiency, and prehistoric harvest pressure: Ichthyofaunal evidence from the Emeryville Shellmound, California. Antiquity 71:845–862. https://doi.org/10.1017/S0003598X0008577X

Broughton JM, O'Connell JF (1999) On evolutionary ecology, selectionist archaeology and behavioral archaeology. Am Antiq 64:153–165. https://doi.org/10.2307/2694351

Bujalesky GG (2011) La inundación del Valle Beagle (11.000 AÑOS A.P.), Tierra del Fuego. The Flood of the Beagle Valley (11.000 YR B.P.), Tierra del Fuego. Anal Inst Patag 39(1):5–21. https://www.analesdelinstitutodelapatagonia.cl/index.php/analespatagonia/article/view/225

Caminos RL, Nullo FF (1979) Descripción geológica de la Hoja 67e, Isla de los Estados. Serv Geológ Nac, Buenos Aires

Carlson D (2022) Quantitative methods in archaeology using R. Cambridge Manuals in Archaeology. Cambridge University Press, Cambridge

Ceraso A, Tessone A, Vázquez MM, Zangrando AF (2010) Recientes prospecciones en Bahía Buen Suceso y Caletas Sorpresa, Herradura y San Mauricio (Península Mitre, Tierra del Fuego). In: Bárcena JR, Chiavazza H (eds) Arqueología Argentina en el Bicentenario de la Revolución de Mayo. XVII Congreso Nacional de Arqueología Argentina. FFyL, UNCUYO, Mendoza, pp 1883–1888

Chapman A (1987) Isla de los Estados en la Prehistoria: Primeros Datos Arqueológicos. Eudeba, Buenos Aires

Christensen M (2016) La industria ósea de los cazadores–recolectores: el caso de los nómadas marinos de Patagonia y Tierra del Fuego. Ediciones Universidad de Magallanes, Punta Arenas

Collado, L (2001) Los bosques de Tierra del Fuego. Análisis de su estratificación mediante imágenes satelitales para el inventario forestal de la provincia. Multequina 10(1):1–16

Corbat M, Tivoli AM, Zangrando AF (2020) Aprovisionamiento de peces en el extremo oriental del canal Beagle (Tierra del Fuego, Argentina): una evaluación a partir del registro zooarqueológico de Bahía Moat. Archaeofauna 29:41–57. https://doi.org/10.15366/archaeofauna2020.29.003

Coronato A (2014) Territorios fueguinos: fisonomía, origen, evolución. In: Oría J, Tivoli AM (eds) Cazadores de mar y tierra. Estudios recientes en arqueología fueguina. Editora Cultural Tierra del Fuego & Museo del Fin del Mundo, Ushuaia, pp 43–63

Coronato A, Borromei AM, Rabassa J (2007) Paleoclimas y Paleoescenarios en la Patagonia Austral y en Tierra del Fuego durante el Cuaternario. Bol Geográf 30:23–38. https://revele.uncoma.edu.ar/index.php/geografia/article/view/100
Couve E, Vidal C (2003) Birds of Patagonia, Tierra del Fuego and Antarctic Peninsula. Ed. Fantástico Sur Birding Ltda, Punta Arenas
Dudley TR, Crow GE (1983) A contribution to the flora and vegetation of Isla de los Estados (Staten Island), Tierra del Fuego, Argentina. Antarct Res Ser 37. American Geophysical Union, Washington DC. https://doi.org/10.1029/AR037
Eidshaug JSP, Bjerck HB, Breivik HM, Risbøl O, Tivoli AM, Zangrando AFJ (2024) From earth to sky: Large-scale archaeological settlement patterns in southernmost South America based on ground surveys, UAV LiDAR, and open access satellite imagery. J Island Coast Archaeol 1-40. https://doi.org/10.1080/15564894.2024.2426828
Erlandson J (2012) A Land by the sea: an ocean view of California archaeology. In: Jones TL, Perry JE (eds) Contemporary issues in California archaeology. Left Coast Press Inc., California, pp 21–35
Estevez J, Piana E, Schiavini A, Juan-Muns N (2001) Archaeological Analysis of Shell Middens in the Beagle Channel, Tierra del Fuego Island. Int J Osteoarch 11:24–33. https://doi.org/10.1002/oa.543
Estévez J, Martínez J (1995) Reflexiones metodológicas a partir del análisis de los restos de pinnípedos de Túnel VII. In: Estévez Escalera J, Vila Mitjá A (eds) Encuentros en los conchales fueguinos, Treballs d´Etnoarqueología 1. CSIC/Universidad Autónoma de Barcelona, Barcelona, pp 183–193
Evans S, Briz I, Álvarez MR, Rowsell K, Collier P, Prosser de Goodall RN, Mulville J, Lacrouts A, Collins MJ, Speller C (2016) Using combined biomolecular methods to explore whale exploitation and social aggregation in hunter–gatherer–fisher society in Tierra del Fuego. J Archaeol Sci Rep 6:757–767. https://doi.org/10.1016/j.jasrep.2015.10.025
Figuerero Torres MJ, Mengoni Goñalons G (1986) Excavaciones arqueológicas en la isla El Salmón (Parque Nacional de Tierra del Fuego), Informes de Investigación 4. Programas de Estudios Prehistóricos–CONICET, Buenos Aires
Fiore D (2011) Art in time. Diachronic rates of change in the decoration of bone artefacts from the Beagle Channel region (Tierra del Fuego, Southern South America). J Anthropol Archaeol 30:484–501. https://doi.org/10.1016/j.jaa.2011.07.002
Fiore D (2006) Puentes de agua para el arte mobiliar: la distribución espacio–temporal de artefactos óseos decorados en Patagonia meridional y Tierra del Fuego. Cazadores–Recolectores del Cono Sur. Rev Arqueol I:137–147
Fiore D (2012) Diseños y *tempos* en el arte mobiliar del canal Beagle (Tierra del Fuego). Una exploración de los ritmos de cambio en la decoración de artefactos óseos. Rel Soc Arg Antropol XXXVII(1):183–206
Flores CE, Deferrari G, Collado L, Escobar J, Schiavini A (2018) Spatial abundance models and seasonal distribution for guanaco (*Lama guanicoe*) in central Tierra del Fuego. Argentina. Plos ONE 13(5):e0197814. https://doi.org/10.1371/journal.pone.0197814
Gifford-González D (2018) An Introduction to zooarchaeology. Springer, Cham
Goodall RNP (1978) Report on the small cetaceans stranded on the coasts of Tierra del Fuego. Sci Rep Whales Res Inst 30:197–230
Goodall N, Dellabianca N, Boy C, Benegas L, Pimper L, Riccialdelli L (2008) Review of small cetaceans stranded or incidentally captured on the coasts of Tierra del Fuego, Argentina, over 33 years. IWC, Sci Comm Doc SC/60/SM21
Gordillo S, Bujalesky G, Pirazzoli P, Rabassa J, Saliège J-F (1992) Holocene raised beaches along the northern coast of the Beagle Channel, Tierra del Fuego, Argentina. Palaeogeogr Palaeoclimatol Palaeoecol 99:41–54. https://doi.org/10.1016/0031-0182(92)90006-Q
Grayson DK (1984) Quantitative zooarchaeology. Academic Press, Orlando
Horwitz VD, Weissel M (2011) Arqueología de Isla de los Estados: la frontera de la abundancia. In: Zangrando AF, Vázquez M, Tessone A (eds) Los cazadores–recolectores del extremo

oriental fueguino. Arqueología de Península Mitre e Isla de los Estados. Sociedad Argentina de Antropología, Buenos Aires, pp 271–286
Horwitz VD (1990) Maritime settlement patterns in southeastern Tierra del Fuego (Argentina). PhD Dissertation, University of Kentucky
Iturraspe RJ, Schroeder C (1999) El clima en el canal Beagle. In: Orquera LA, Piana EL (eds) La vida material y social de los Yámana, EUDEBA–IFIC, Buenos Aires, pp 36–45
Juan–Muns i Plans N (1992) La pesca com alternativa per als Yàmana, nòmades canoers del Canal Beagle (Tierra del Fuego, Argentina). PhD Dissertation, Universidad Autónoma de Barcelona
Juan–Muns i Plans N (1996) Aprovechamiento de recursos ícticos en Túnel VII (Tierra del Fuego). In: Gómez Otero J (ed) Arqueología. Sólo Patagonia. CENPAT, Puerto Madryn, pp 89–97
Kochi S, Pérez SA, Tessone A, Ugan A, Tafuri MA, Nye J, Tivoli AM, Zangrando AF (2018) $\delta^{13}C$ and $\delta^{15}N$ variations in terrestrial and marine foodwebs of Beagle Channel in the Holocene. Implications for human paleodietary reconstructions. J Archaeol Sci Rep 18:696–707. https://doi.org/10.1016/j.jasrep.2017.11.036
Lanata JL (1996) The "Haush" puzzle: piecing together subsistence and settlement at the fueguian southeast. Rev Mus Arqueol Etnol 6:11–32. https://doi.org/10.11606/issn.2448-1750.revmae.1996.109251
Legoupil D (1993–94) El archipiélago del Cabo de Hornos y la costa sur de la isla Navarino: poblamiento y modelos económicos. Anal Instit Patag 22:101–121
Legoupil D, Fontugne M (1997) El poblamiento marítimo en los archipiélagos de Patagonia: núcleos antiguos y dispersión reciente. Anal Inst Patag 25:75–87
Lloris D, Rucabado J (1991) Ictiofauna del Canal Beagle (Tierra del Fuego), aspectos ecológicos y análisis biogeográfico. Inst Esp Oceanog (Public Esp 8), Madrid. http://hdl.handle.net/10261/198508
López HL, García M, San Román N (1996) Lista comentada de la ictiofauna del Canal Beagle, Tierra del Fuego, Argentina, Contribución Científica (Publicación Especial). CADIC, Ushuaia
Lovrich G (2014) Lithodidae. In: Calcagno JA (ed) Los Invertebrados del Mar Argentino. Vazquez Mazzini Editores, Buenos Aires
Lyman L (1994) Vertebrate taphonomy. Cambridge University Press, Cambridge
Lyman L (2003) The influence of time averaging and space averaging on the application of foraging theory in zooarchaeology. J Archaeol Sci 30(595):610. https://doi.org/10.1016/S0305-4403(02)00236-4
Mameli L (2000) Análisis arqueoavifaunístico del yacimiento Túnel VII, Tierra del Fuego, Argentina. Master Dissertation, Universidad Autónoma de Barcelona
Mameli L (2003) La gestión del recurso avifaunístico por las poblaciones canoeras del archipiélago fueguino. PhD Dissertation, Universidad Autónoma de Barcelona
Mameli L, Estévez Escalera J (2004) Etnoarqueozoología de aves: el ejemplo del extremo sur americano, Treballs D'Etnoarqueología, 5. Universidad Autónoma de Barcelona y Consejo Superior de Investigaciones Científicas CSIC, Madrid
Martinoli MP (2017) Pautas de procesamiento y consumo de pinnípedos en la costa sur de Tierra del Fuego e Isla de los Estados: un análisis comparativo. Arqueología 23(3):173–196. http://revistascientificas.filo.uba.ar/index.php/Arqueologia/article/view/4013
Martinoli MP, Vázquez M (2017) Pinniped Capture and Processing: A Comparative Analysis from Beagle Channel (Tierra del Fuego, Argentina). In: Mondini M, Muñoz AS, Fernández PM (eds) Zooarchaeology in the neotropics. Environmental diversity and human–animal interactions. Springer, Cham, pp 7–23
Martinoli MP (2018) Modalidades de explotación, procesamiento y consumo de pinnípedos en la margen meridional de Tierra del Fuego. PhD Dissertation, Universidad de Buenos Aires
Martinoli MP (2019) La explotación de pinnípedos en Isla de los Estados: el caso de las ocupaciones de Bahía Crossley I. Rev Mus Antropol 12(3):43–60. https://doi.org/10.31048/1852.4826.v12.n3.24514

Martinoli MP, Zangrando AF (2024) Estrategias de aprovisionamiento de otáridos en la margen meridional de Tierra del Fuego durante el Holoceno. Lat Am Antiq 1–18. https://doi.org/10.1017/laq.2023.60

Martinoli MP, Crespo E, Dellabianca N, Zangrando AF (2021) Nuevos datos de estimación de edad y estación de muerte de pinnípedos en la región del canal Beagle, Tierra del Fuego (Argentina). Intersecc Antropol 22(2):195–207. https://doi.org/10.37176/iea.22.2.2021.629

Martinoli MP, Zangrando AF, Tivoli AM, Paskulin L, Hilsden J, Williamson D, Evans Z, Speller C (2022) Los restos de cetáceos en el canal Beagle: nuevos resultados a partir de Zooarqueología por espectrometría de masas (ZooMS). Paper presented at the 6to Congreso Nacional de Zooarqueología Argentina, University of La Plata, La Plata, 14–18 November 2022

Mengoni Goñalons G (1988) Análisis de materiales faunísticos de sitios arqueológicos. Xama I:71–120

Mengoni Goñalons G (1999) Cazadores de guanacos de la estepa patagónica. Sociedad Argentina de Antropología, Buenos Aires

Milano VN, Grandi MF, Schiavini A, Crespo E (2020a) Sea lions (*Otaria flavescens*) from the end of the world: insights of a recovery. Polar Biol 43:695–706. https://doi.org/10.1007/s00300-020-02672-9

Milano VN, Grandi MF, Schiavini A, Crespo E (2020b) Recovery of South American fur seals from Fuegian Archipelago (Argentina). Mar Mammal Sci 1–11. https://doi.org/10.1111/mms.12686

Moraga CA, Funes MC, Pizarro JC, Briceño C, Novaro AJ (2015) Effects of livestock on guanaco *Lama guanicoe* density, movements and habitat selection in a forest–grassland mosaic in Tierra del Fuego, Chile. Oryx 49(1):30–41. https://doi.org/10.1017/S0030605312001238

Nagaoka L (2005) Differential recovery of Pacific Island fish remains. J Archaeol Sci 32:941–955. https://doi.org/10.1016/j.jas.2004.12.011

Narosky T, Izurieta D (2003) Aves de Argentina y Uruguay: guía para la identificación. Edición de oro. Ed. Vázquez Mazzini, Buenos Aires

Nye J, Zangrando AF, Martinoli MP, Fogel M (2020) Temporal and population trends in human exploited pinnipeds from Tierra del Fuego. Palaeogeogr Palaeoclimatol Palaeoecol 554:109804. https://doi.org/10.1016/j.palaeo.2020.109804

Ocampo C, Rivas P (2000) Nuevos fechados ^{14}C de la costa norte de la Isla Navarino, costa sur del canal Beagle, Provincia Antártica Chilena, Región de Magallanes. Anal Inst Patag. Ser Cs Hs 28:197–214

Ocampo C, Rivas P (2002) Arqueología del canal Beagle: secuencias y procesos culturales en ambientes de altas latitudes. Bol Soc Chil Arqueol 33(34):98–102

Orquera LA, Piana EL (1983) Adaptaciones marítimas prehistóricas en el litoral magallánico fueguino. Rel Soc Arg Antropol XV:225–235

Orquera LA, Piana EL (1988) Human littoral adaptation in the Beagle Channel region: the maximun possible age. Quat S Am Antarc Penin 5:133–165

Orquera LA, Piana EL (1989–90) La formación de los montículos arqueológicos de la región del canal Beagle. Runa XIX:59–82. https://doi.org/10.34096/runa.v19i1.4271

Orquera LA, Piana EL (1993–94) Lancha Packewaia: actualización y rectificaciones. Rel Soc Arg Antropol XIX:325–362

Orquera LA, Piana E (1995) Túnel VII en la secuencia arqueológica del canal Beagle: hipótesis y expectativas de los investigadores argentinos. In: Estévez Escalera J, Vila Mitjá A (eds) Encuentros en los conchales fueguinos, Treballs d'Etnoarqueología 1. CSIC/Universidad Autónoma de Barcelona, Barcelona, pp 25–45

Orquera LA, Piana EL (1996) El sitio Shamakush I (Tierra del Fuego, República Argentina). Rel Soc Arg Antropol XXI:215–265

Orquera LA, Piana EL (1999a) Arqueología de la región del canal Beagle (Tierra del Fuego, República Argentina). Sociedad Argentina de Antropología, Buenos Aires

Orquera LA, Piana EL (1999b) La vida material y social de los Yámana. Eudeba, IFIC, Buenos Aires

Orquera LA, Piana E (2000) Imiwaia I: un sitio de canoeros del sexto milenio A.P. en la costa norte del canal Beagle. In: Desde el País de los Gigantes. Perspectivas arqueológicas en Patagonia II, Universidad Nacional de la Patagonia Austral, Río Gallegos, pp 441–453

Orquera LA, Piana EL (2005) La adaptación al litoral sudamericano sudoccidental: qué es y quiénes, cuándo y dónde se adaptaron. Rel Soc Arg Antropol 30:11–32

Orquera LA, Piana EL (2006) El poblamiento inicial del área litoral sudamericana sudoccidental. Magallania 34(2):21–36. https://doi.org/10.4067/S0718-22442006000200003

Orquera LA, Piana EL (2009) Sea nomads of the Beagle Channel in southernmost South America: over six thousand years of coastal adaptation and stability. J Island Coast Archaeol 4:61–81. https://doi.org/10.1080/15564890902789882

Orquera L, Legoupil D, Piana E (2011) Littoral adaptation at the southern end of South America. Quat Int 239(1):61–69. https://doi.org/10.1016/j.quaint.2011.02.032

Ortiz Troncoso O (1975) Los yacimientos de Punta Santa Ana y Bahía Buena (Patagonia Austral). Excavaciones y fechados radiocarbónicos. Anal Inst Patag VII(1–2):93–122

Piana EL (1984) Arrinconamiento o adaptación en Tierra del Fuego. In: Antropología Argentina 1984. Editorial de Belgrano, Buenos Aires, pp 7–114

Piana EL (2005) Cetaceans and human beings at the uttermost part of America: a lasting relationship in Tierra del Fuego. In: Monks G (ed) The exploitation and cultural importance of sea mammals. Oxbow Books, Oxford, pp 121–137

Piana EL, Canale G (1993–94) Túnel II: Un yacimiento de la Fase Reciente del canal Beagle. Rel Soc Arg Antropol XIX:363–389

Piana EL, Orquera LA (2007) Diferencias regionales y temporales en el litoral sudoccidental de Sudamérica. In: Morelo F, Martinic M, Prieto A, Bahamonde G (eds) Arqueología de Fuego-Patagonia. Levantado piedras, desenterrando huesos... y develando arcanos. Ediciones CEQUA, Punta Arenas, pp 311–323

Piana EL, Vázquez MM (2008) El Sitio Shamakush VIII: puntualizaciones sobre el uso de recursos y la gestión del asentamiento en el canal Beagle. In: Austral A, Tamagnini M (eds) Problemáticas de la Arqueología Contemporánea, vol III. Universidad Nacional de Río Cuarto, Río Cuarto, pp 1057–1068

Piana EL, Estévez Escalera J, Vila Mitjá A (2000) Lanashuaia: un sitio de canoeros del siglo pasado en la costa norte del canal Beagle. In: Belardi JB, Carballo Marina F, Espinosa S, De Nigris M, Fernández P, Gómez Otero J, Goñi R (eds) Desde el País de los Gigantes. Perspectivas arqueológicas en Patagonia, vol II. Universidad Nacional de la Patagonia Austral, Río Gallegos, pp 455–469

Piana E, Vázquez MM, Rúa N (2004) Mischiuén I. Primeros resultados de una excavación de rescate en la costa norte del canal Beagle. In: Civalero M, Fernández P, Guráieb G (eds) Contra viento y marea. Arqueología de la Patagonia. INAPL, Buenos Aires, pp 815–832

Piana EL, Vázquez M, Tivoli AM (2007a) Dieta y algo más. Animales pequeños y variabilidad en el comportamiento humano en el canal Beagle. In: Arqueología de Fuego–Patagonia. Levantando piedras, desenterrando huesos… y develando arcanos. Ediciones CEQUA, Punta Arenas, pp 39–50

Piana EL, Vázquez MM, Ceraso A (2007b) Kaiyawoteha. Informe de excavación. MS

Piana EL, Vázquez M, Álvarez M (2008) Nuevos resultados del estudio del sitio Ajej I: un aporte a la variabilidad de estrategias de los canoeros fueguinos. Runa 29:87–100

Piana EL, Zangrando AF, Orquera LA (2012) Early occupations in Tierra del Fuego and the Evidence from Layer S at the Imiwaia I Site (Beagle Channel, Argentina). In: Miotti L, Salemme M, Flegenheimer N, Goebel T (eds) Southbond. Late Pleistocene Peopling of Latin America. Special Edition, Current Research in the Pleistocene. Center for the Study of the First Americans, Texas, pp 171–175

Ponce J (2009) Palinología y geomorfología del Cenozoico tardío de la Isla de los Estados. PhD Dissertation, Universidad Nacional del Sur

Ponce J, Fernández M (2014) Climatic and Environmental History of Isla de los Estados, Argentina. Springer, Dordrecht, Heidelberg, New York, London. https://doi.org/10.1007/978-94-007-4363-2

Ponce JF, Borromei AM, Rabassa JO (2011) Evolución del paisaje y de la vegetación durante el cenozoico tardío en el extremo sureste del Archipiélago fueguino y el canal Beagle. In: Zangrando AF, Vázquez M, Tessone A (comps) Los cazadores–recolectores del extremo oriental fueguino. Arqueología de Península Mitre e Isla de los Estados. Sociedad Argentina de Antropología, Buenos Aires, pp 31–64

Rabassa J, Coronato A, Bujalesky G, Salemme M, Roig C, Meglioli A, Heusser C, Gordillo S, Roig F, Borromei A, Quattrocchio M (2000) Quaternary of Tierra del Fuego, southernmost South America: an updated review. Quat Int 68–71:217–240. https://doi.org/10.1016/S1040-6182(00)00046-X

Raya Rey A, Schiavini A (2000) Distribution, abundance and associations of seabirds in the Beagle Channel, Tierra del Fuego, Argentina. Polar Biol 23:338–345. https://doi.org/10.1007/s00300 0050453

Reitz EJ, Wing ES (1999) Zooarchaeology. In: Cambridge manuals in archaeology. Cambridge University Press, Cambridge

Roig C, Collado L (2004) Los turbales patagónicos, ventanas N° 6 & 7. In: Blanco D, de la Balze V (eds) Los turbales de la Patagonia. Bases para su inventario y la conservación de su biodiversidad 19. Wetlands International, Buenos Aires, pp 62–71

San Román, M (2013) Sitios arqueológicos de Isla Englefield, Mar de Otway: nuevas evidencias de discontinuidad cultural en el proceso de poblamiento marítimo de Patagonia meridional. In: Zangrando AF, Barberena R, Gil A, Neme G, Giardina M, Luna L, Otaola C, Paulides S, Salgán L, Tivoli AM (eds) Tendencias Teórico Metodológicas y Casos de Estudio en la Arqueología de Patagonia. Sociedad Argentina de Antropología, INAPL, Museo de Historia Natural de San Rafael, Buenos Aires, pp 523–534

Santiago F, Vázquez M (2012) Dietas promediadas: explorando el registro zooarqueológico supra–regional en Tierra del Fuego. Rev Mus Antropol 5:225–238. https://doi.org/10.31048/1852.4826.v5.n1.9193

Scheinsohn V (1997) Explotación de materias primas óseas en la Isla Grande de Tierra del Fuego. PhD Dissertation, Universidad de Buenos Aires

Scheinsohn V (2010) Hearts and Bones. bone raw material exploitation in Tierra del Fuego. BAR International Series 2094, Oxford. https://doi.org/10.30861/9781407305707

Schiavini A (1993) Los lobos marinos como recurso para cazadores–recolectores marinos: El caso de Tierra del Fuego. Lat Am Antiq 4(4):346–366. https://doi.org/10.2307/972072

Schiavini A, Yorio P (1995) Distribution and abundance of seabirds colonies in the Argentine sector of the Beagle Channel (Tierra del Fuego). Mar Ornithol 23:39–46

Schiavini A, Raya Rey A (2001) Aves y mamíferos marinos en Tierra del Fuego. Estado de situación, interacción con actividades humanas y recomendaciones para su manejo. Informe preparado bajo contrato con el Proyecto Consolidación e Implementación del Plan de Manejo de la Zona Costera Patagónica, Proyecto ARG/97/G31 GEF/PNUD/MRECIC, CADIC–CONICET, Ushuaia

Schiavini A (1990) Estudio de la relación entre el hombre y los pinnípedos en el proceso adaptativo humano al canal Beagle, Tierra del Fuego, Argentina. PhD Dissertation, Universidad de Buenos Aires

Sielfeld W (1983) Mamíferos Marinos de Chile. Universidad de Chile, Santiago

Smith AB, Kinahan J (1984) The invisible whale. World Archaeol 16(1):89–97. https://doi.org/10.1080/00438243.1984.9979918

Thomas DH (1969) Great basin hunting patterns: a quantitative method for treating faunal remains. Ame Antiq 34:392–401. https://doi.org/10.2307/277736

Tivoli AM (2010a) Exploitation of bird resources among prehistoric sea–nomad societies of the Beagle Channel region, southern South America. Before Farming (2):1–12. https://doi.org/10.3828/bfarm.2010.2.3

Tivoli AM (2010b) Temporal trends in avifaunal resource management by prehistoric sea nomads of the Beagle Channel region (southern South America). In: Prummel W, Zeiler JT, Brinkhuizen DC (eds) Birds in archaeology. Proceedings of the 6th meeting of the ICAZ Bird working group in groningen, groningen archaeological studies, vol 12. Barkhuis Groningen University Library, Groningen, pp 131–140

Tivoli AM (2012) ¿Intensificación? en el aprovechamiento de aves entre los cazadores–recolectores–pescadores de la región del canal Beagle. Archaeofauna 21:121–137. https://doi.org/10.15366/archaeofauna2012.21.006

Tivoli AM (2014) Processing activities and differentiation of bird utilization during the Late Holocene in the Beagle Channel region (Southern South America). Int J Osteoarchaeol 24:397–406. https://doi.org/10.1002/oa.2374

Tivoli AM, Zangrando AF (2011) Subsistence variations and landscape use among maritime hunter–gatherers. A zooarchaeological analysis from the Beagle Channel (Tierra del Fuego, Argentina). J Archaeol Sci 38(5):1148–1156. https://doi.org/10.1016/j.jas.2010.12.018

Tivoli AM, Montes A, Alunni DV, Fernández Ropero MC, Martinoli MP, Pérez SA, Pinto Vargas G, López R, Saenz Samaniego R, Zangrando AF (2022) Arqueología de sectores insulares en el canal Beagle. El caso del archipiélago de las Islas Bridges e islas adyacentes. Arqueología 28(1):9976. https://doi.org/10.34096/arqueologia.t28.n1.9976

Tivoli AM, Barrientos Contreras MJ, Zangrando AF (2025) Human use of bird resources in the northeastern coasts of the Beagle Channel (Tierra del Fuego, southern South America) during the Late Holocene. Int J Osteoarchaeol 1–14. https://doi.org/10.1002/oa.70041

Vaz Ferreira R, Vallejo S (1980) Algunos aspectos del comportamiento de *Arctocephalus australis* (Zimmerman) lobo de dos pelos de Sudamérica (pinnipedia–otariidae) en el Uruguay. In: Actas de las VI Jornadas Argentinas de Zoología. Museo de La Plata, La Plata, pp 223–236

Vaz Ferreira R (1981) South American sea lion *Otaria flavescens* (Shaw 1800). In: Ridgway S, Harrison R (eds) Handbook of marine mammals. Academic Press, Cambridge, pp 39–65

Vázquez MM (2015) Guanacos en el segundo componente de Túnel I (canal Beagle). Un enfoque tafonómico. Magallania 43(1):251–277. https://doi.org/10.4067/S0718-22442015000100014

Vázquez MM (2024) Movilidad y subsistencia de los cazadores–recolectores de Península Mitre, extremo sudeste de Tierra del Fuego. PhD Dissertation, Universidad de Buenos Aires

Vázquez M, Prieto A (2014) Estudio preliminar. Búsqueda sin término: breve historia de la arqueología en Tierra del Fuego. In: Oría J, Tivoli A (eds) Cazadores de mar y tierra. Estudios recientes en arqueología fueguina. Editora Cultural Tierra del Fuego & Museo del Fin del Mundo, Ushuaia, pp 15–42

Vázquez M, Zangrando AF (2017) Estructuras de pesca en el canal Beagle. Magallania (Punta Arenas) 45(1):101–122. https://doi.org/10.4067/S0718-22442017000100101

Vázquez M, Borrero L (2021) Sinopsis de la arqueología de Tierra del Fuego. Rev Esp Antropol Am 51:173–185. https://doi.org/10.5209/reaa.72827

Vázquez MM, Zangrando AF, Tessone A, Ceraso A, Sosa L (2007) Arqueología de Bahía Valentín (Península Mitre, Tierra del Fuego): nuevos resultados y perspectivas. In: Morello F, Martinic M, Prieto A, Bahamonde G (eds) Arqueología de Fuego–Patagonia. Levantando piedras, desenterrando huesos...y develando arcanos. Ediciones CEQUA, Punta Arenas, pp 755–766

Vázquez MM, Zangrando AF, Tessone A, Ceraso A (2011) Arqueología de la costa meridional de Península Mitre. In: Zangrando AF, Vázquez MM, Tessone A (comps) Los cazadores–recolectores del extremo oriental fueguino. Arqueología de Península Mitre e Isla de los Estados. Sociedad Argentina de Antropología, Buenos Aires, pp 203–230

Vázquez M, Martinoli MP, Tivoli AM, Pinto Vargas G, Piana EL, Alunni D, Zangrando AF (2018) Kaiyawoteha II. Pautas de aprovisionamiento de recursos animales en la bahía de Ushuaia durante el Holoceno tardío. Paper presented in: IV Encuentro Latinoamericano de ArqueoZoología, (IV ELAZ), Río Gallegos, 21–23 November 2018

Vidal H (1985) Los conchales de Bahía Valentín. Degree Dissertation, Universidad de Buenos Aires

Vidal H (1988) Bahía Valentín: 6000 años de ocupaciones humanas en el oriente fueguino. In: Libro de resúmenes del IX Congreso Nacional de Arqueología Argentina. FFyL, Universidad de Buenos Aires, Buenos Aires, p 77

Wheeler A, Jones AKG (1989) Fishes. Cambridge University Press, Cambridge

Winterhalder B, Baillargeon W, Cappelletto F, Daniel I, Prescott C (1988) The population ecology of hunter–gatherers and their prey. J Anthropol Archaeol 7(4):289–328. https://doi.org/10.1016/0278-4165(88)90001-3

Zangrando AF (2003) Ictioarqueología del canal Beagle: Explotación de peces y su implicación en la subsistencia humana. Sociedad Argentina de Antropología, Buenos Aires

Zangrando AF (2009a) Historia evolutiva y subsistencia de cazadores–recolectores marítimos de Tierra del Fuego. Sociedad Argentina de Antropología, Buenos Aires

Zangrando AF (2009b) Is fishing intensification a direct route to hunter–gatherer complexity? A case study from the Beagle Channel region (Tierra del Fuego, southern South America). World Archaeol 41(4):589–608. https://doi.org/10.1080/00438240903363848

Zangrando AF (2010) Coastal archaeology and Hunter-Gatherers in Southeastern Tierra del Fuego. J Island Coast Archaeol 5(2):288–291. https://doi.org/10.1080/15564894.2010.487360

Zangrando AF (2018) Poblamiento temprano y arqueología de costas en Patagonia y Tierra del Fuego: vacío de información, preconceptos y perspectivas. Foro de discusión. Intersecc Antropol 19:63–97

Zangrando AF, Tessone A, Vázquez MM (2009) El uso de espacios marginales en el archipiélago fueguino: implicaciones de la evidencia arqueológica de Bahía Valentín. In: Salemme M, Santiago F, Álvarez MR, Piana EL, Vázquez MM, Mansur E (eds) Arqueología de Patagonia: una mirada desde el último confín. Editorial Utopías, Ushuaia, pp 47–62

Zangrando AF, Alunni DV, Martinoli MP, Tivoli AM, Piana EL (2010) Arqueología de la región de Moat (Tierra del Fuego, Argentina): estudios preliminares en la localidad arqueológica Heshkaia. In: Bárcena J, Chiavazza H (eds) Arqueología Argentina en el Bicentenario de la Revolución de Mayo, Actas XVII Congreso Nacional de Arqueología Argentina. FFyL, UNCUYO, Mendoza, pp 2005–2010

Zangrando AF, Borrazzo K, Tivoli AM, Alunni DV, Martinoli MP (2014) El sitio Heshkaia 35: nuevos datos sobre la arqueología de Moat (Tierra del Fuego, Argentina). Rev Mus Antropol 7(1):11–24. https://doi.org/10.31048/1852.4826.v7.n1.9090

Zangrando AF, Tivoli AM (2015) Human use of birds and fish in marine settings of southern Patagonia and Tierra del Fuego in the Holocene: a first macro–regional approach. Quat Int 373:82–95. https://doi.org/10.1016/j.quaint.2014.11.047

Zangrando AF, Vázquez M, Tessone A (eds) (2011) Los cazadores-recolectores del extremo oriental fueguino Arqueología de Península Mitre e Isla de los Estados. Sociedad Argentina de Antropología, Buenos Aires

Zangrando AFJ, Ponce JF, Martinoli MP, Montes A, Piana E, Vanella F (2016) Palaeogeographic changes drove prehistoric fishing practices in the Cambaceres Bay (Tierra del Fuego, Argentina) during the middle and late Holocene. Environmental Archaeology 21(2): 182–192. http://dx.doi.org/10.1080/14614103.2015.1130888

Zangrando AF, Bjerck HB, Piana EL, Breivik HM, Tivoli AM, Negre J (2018) Spatial patterning and occupation dynamics during the Early Holocene in an archaeological site from the south coast of Tierra del Fuego: Binushmuka I. Estud. Atacameños 60:31–49. https://doi.org/10.4067/S0718-10432018005001401

Zangrando AF, Tivoli AM, Fernández Ropero C, Pinto Vargas G (2021a) Contextos arqueológicos en paleocostas: un nuevo caso de estudio en Bahía Moat (Tierra del Fuego). Paper presented at the XI Jornadas de Arqueología de la Patagonia 2021, Universidad Austral de Chile, Puerto Montt, 29 November–3 December 2021

Zangrando AF, Tivoli AM, Ponce JF, Alunni DV, Fernández Ropero C, Martinoli MP, Montes A (2021b) Colonización humana de Tierra del Fuego: contextos arqueológicos pre-transgresivos en la costa norte del canal Beagle. Chungará (Arica) 54:557–577

Zangrando AF, Kochi S, Nye J, Tessone A, Riccialdelli L, Tivoli AM, Martinoli MP, Pérez SA, Fogel M (2022) Long–term interaction between Hunter–Gatherer–fisher societies and marine resources in the Southern tip of South America applications of stable isotopes in zooarchaeology. In: Sharpe A, Krigbaum J (eds) Isotope research in zooarchaeology: methods, applications, and advances. University Press of Florida, Florida, pp 15–43. https://ri.conicet.gov.ar/handle/11336/210353

Zangrando AF, Martinoli MP, Tessone A, Vázquez MM, Alunni DV, Franch Bach A, Hernandez Herrero O, Fernández Ropero C, Pinto Vargas G, Tivoli AM (2024a) El uso de espacios marginales en el archipiélago de Tierra del Fuego: el caso de Isla de los Estados. Bol Soc Chil Arqueol 57:85–119. https://doi.org/10.56575/BSCHA.05700240843

Zangrando AF, Tivoli AM, Martinoli MP, Christensen M, Paskulin L, Hilsden J, Loewen D, Evans Z, Speller C (2024b) ZooMS as a tool to explore bone technology production. A case study from Tierra del Fuego (southern South America). In: Paper presented at the 15th International Meeting of the Worked Bone Research Group, ICAZ, Paris, 13–17 May 2024

Zohar I, Belmaker M (2005) Size does matter: methodological comments on sieve size species richness in fishbone assemblages. J Archaeol Sci 32:635–641. https://doi.org/10.1016/S0305-4403(03)00037-2

Chapter 4
Mussel Sizes in Shell Middens: Insights into the Long-Term Human Harvesting of Intertidal Shellfish in Southern Tierra del Fuego

Suray A. Pérez, Angélica M. Tivoli, and Atilio Francisco J. Zangrando

Abstract In this chapter, we conduct a comprehensive review of dimensional data related to mussel shells retrieved from archaeological shell middens from different coastal localities in southern Tierra del Fuego. In this area, human procurement of shellfish predominantly targeted the naturally abundant intertidal species, blue mussels (*Mytilus edulis*), during the second half of the Holocene. This bivalve retained its role as the dominant species throughout the entire archaeological sequence, showing non-directional shifts in mean shell size over time. Previous studies interpreted fluctuations in mean valve length as indicative of localized impacts on mussel populations, suggesting a sustainable harvesting strategy by highly mobile groups periodically exploiting alternative coastal areas. However, recent zooarchaeological and palaeodemographic analyses suggest a more intensive use of littoral zones during the Late Holocene along the southern coast of Tierra del Fuego. This prompts new inquiries into the temporal dynamics of shellfish gathering practices, hinting at the possibility that local mussel subpopulations faced variable harvesting intensity as part of an overarching process of subsistence changes over time. Diachronic trends in shell sizes are then examined to initially assess the scope and limitations of metrical information for tracking long-term changes in human predation pressure on mollusks within a general context of sustainable use. Additionally, by considering the ecological and actualistic information of these bivalves, their implications are evaluated in light of the area's occupational history.

A. M. Tivoli · A. F. J. Zangrando
CADIC-CONICET, Ushuaia, TDFeIAS, Argentina
e-mail: tivoli.angelica@conicet.gov.ar

A. F. J. Zangrando
e-mail: fzangrando@conicet.gov.ar

S. A. Pérez (✉) · A. F. J. Zangrando
FFyL-UBA, CABA, Argentina
e-mail: suryviaja@filo.uba.ar

A. F. J. Zangrando et al. (eds.), *Archaeology of the South Coast of Tierra del Fuego*, The Latin American Studies Book Series, https://doi.org/10.1007/978-3-032-09487-2_4

4.1 Introduction

Human predation on intertidal shellfish is an important issue in archeological studies focusing on coastal subsistence and shell midden formation during the Holocene (Waselkov 1987; Orquera and Piana 2001; Mannino and Thomas 2002; Parkington and Brand 2020). The scale or intensity of exploitation has often been discussed through optimal foraging models that tend to consider the negative effects of anthropogenic impact documented in several ecological and ethnographic studies (Hockey and Bosman 1986; Lasiak 1992). From this perspective, the consistent removal of larger, higher meat-yielding shellfish species over time will lead inexorably to an irreversible stock depletion (Anderson 1981). The basic archaeological criteria for inferring overexploitation from shell midden samples typically include a reduction in the mean or modal size and age of preferred species over a sequence of deposits spanning centuries or thousands of years, followed by a decrease in their abundance relative to other less easily procured/processed species (Botkin 1980; Anderson 1981; Claassen 1998; Mannino and Thomas 2002). While these features have been recognized worldwide in numerous archaeological settings and ascribed to shellfish resource depression (Botkin 1980; Raab 1992; Jerardino 1997; Mannino and Thomas 2002; Klein et al. 2004; Rick et al. 2006; Braje et al. 2007), recent studies have reported cases where heavily harvested species prevailed over long-term occupational sequences with no signs of stock depression (Erlandson et al. 2008; Hausmann et al. 2021). Such scenarios call attention to other dimensions of predator–prey interactions which are usually dismissed by evolutionary foraging models and could have contributed to sustain a long and persistent shellfish exploitation. For example, certain biological and ecological traits can render resilience to a given species when faced with high predation pressure (Whitaker 2008; Whitaker and Byrd 2014; Codding et al. 2014). Likewise, a combination of some behavioral patterns of prehistoric hunter-gatherers may have helped to ameliorate the depleting effects of human intensive harvesting (Mannino and Thomas 2002). Deliberate collection strategies, including specific methods of procuring shellfish (Whitaker 2008), regulation of harvest intensity (Cannon and Burchell 2009), and intentional cultivation (Groesbeck et al. 2014), along with the periodic movement of residential sites and low human population densities (Mannino and Thomas 2002), have recently been invoked as means of resource management.

On the southern coast of Tierra del Fuego (Fig. 4.1), the ubiquity and abundance of mussels (Mytilidae) among archaeological deposits dating from the last 7500 years points to a long-term stability in predation despite intensive shellfish exploitation (Orquera and Piana 1999; 2001). Changes in shell size over time display permanent oscillations and have been linked to small human impacts on local beds occurring on scales of weeks or months (Orquera and Piana 2001; 2009; Zangrando et al. 2017; Pérez 2020). Formation processes involved in shell-matrix sites suggest they were the outcome of episodic and short-lived human occupations that took place after lengthy gaps of centuries or even millennia, providing the mussel-exploited populations the possibility for recovery (Orquera and Piana 1992, 2009; Zangrando et al.

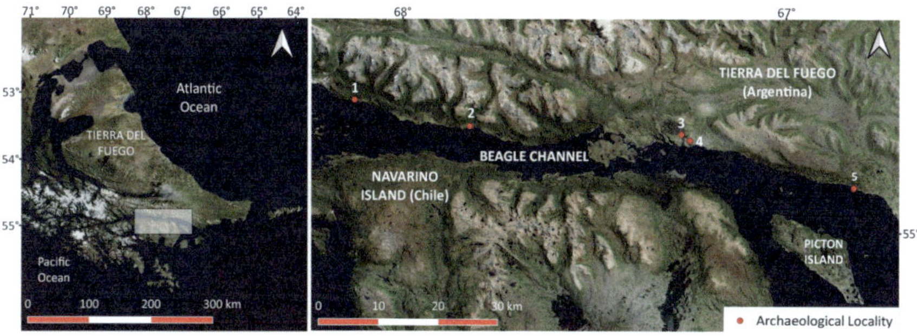

Fig. 4.1 Map the southern coast of Tierra del Fuego showing the location of shell midden sites from different archaeological localities from the Beagle Channel: 1. Túnel-L.Packewaia, 2. Shamakush, 3. Imiwaia, 4. Lanashuaia, 5. Heshkaia

2020). Because mussels are broadly distributed along the south coast of the island, human harvesting would only have affected restricted portions of the distribution area of these mollusks without implying significant risks to their natural sources of replacement (Orquera and Piana 1999: 117, 2009).

Most archaeomalacological data from southern Tierra del Fuego is based on studies conducted in the north coast of Beagle Channel. These prior works supported the idea of continuity and relative stability in the subsistence and demography of hunter-gatherer societies during the Middle and Late Holocene (Orquera and Piana 1999; 2009). Nevertheless, subsequent research in that region and neighboring areas has documented highly diverse archaeological contexts, revealing significant variability in the ways coastal environments and resources were exploited over time (Zangrando 2009; Zangrando et al. 2011; Martinoli 2018; Alunni et al. 2020). In particular, zooarchaeological and palaeodemographic assessments suggest a more intensive use of littoral zones during the Late Holocene along the southern coast of Tierra del Fuego (Zangrando 2009; Orquera and Piana 2020; Zangrando et al. 2022). In this context, new questions arise about the temporal dynamics of shellfish gathering practices in the area. Although a long-term decrease in mussels' natural abundance due to human overexploitation can be ruled out (Orquera and Piana 1999, 2001), it is necessary to investigate whether their local subpopulations experienced increasing levels of harvesting intensity as part of an overall diachronic process of subsistence changes (Zangrando et al. 2011; Zangrando et al. 2022).

Some authors have recognized, however, certain shortcomings when using average shell sizes as a proxy for human harvesting intensity (Claassen 1998; Campbell 2008). It is therefore imperative to assess the scope and limitations of metrical information previously published for tracking changes in human predation pressure within a context of sustained use and continuous renewal of these resources. In this chapter, we conduct a comprehensive review of dimensional data related to mussel shells gathered from archaeological assemblages in various sectors of the southern coast of Tierra del Fuego. Diachronic trends are analyzed, and by taking into account the ecological and actualistic information of these bivalves, their implications are evaluated within the framework of the area's occupational history.

4.2 Shell Middens on the Southern Coast of Tierra del Fuego and Archaeological Expectations

Along the southern coast of Tierra del Fuego shell middens occur as the most ubiquitous and visible archaeological feature (Orquera and Piana 1999; Piana and Orquera 2010; Zangrando 2010). The matrix of these deposits is mainly composed of mollusks remains and fine sediment (Orquera and Piana 2000, 2001; Zangrando et al. 2017). The proportional contribution of shells varies across the stratigraphical sequences of the sites, reflecting different accumulation rates. This variation demonstrates that the processes of their formation were not continuous (Orquera and Piana 1991; Zangrando et al. 2020).

The taxonomic composition of shell assemblages in all middens is fairly uniform with blue mussels *Mytilus edulis platensis* (hereinafter *M. edulis*) comprising the dominant portion. To a lesser extent, *Brachidontes purpuratus* (*B. purpuratus*) and *Aulacomya atra* (*A. atra*) are also present. Overall, bivalves prevail in terms of weight and the minimal number of individuals (MNI) over polyplacophores (*Chiton* sp.) and gastropods, the last including limpets *Nacella sp.* and *Fissurella* sp. and sea snails, such as *Acanthina monodon* and *Trophon geversianus* (Orquera and Piana 2000, 2001; Pérez 2020). Species' relative abundances closely reflect their natural distribution within the intertidal and shallow subtidal zone, suggesting that shellfish availability was the main force shaping collection strategies (Orquera and Piana 2001, 2009; Pérez 2020).

Shell middens are similarly distributed in the area, with over 90% of these structures situated within approximately 300 m from the present coastline (Chap. 2). However, the density of sites decreases toward the eastern end: systematic surveys carried out in Moat Bay indicate two structures per linear kilometer of coastline (Zangrando et al. 2014), while in more westerly sectors located on the northern margin of the Beagle Channel the density is four shell middens per linear kilometer (Barceló et al. 2002). Spatial variation is also evident in the distribution of shell midden types: along the central section of the Beagle Channel up to Cambaceres Bay, ring-shaped mounds are the most common feature and can occur in clusters numbering in the dozens or even hundreds. This feature is not found in Moat Bay, where mounds display variable dimensions and are normally arranged in isolation. Since shell density and size provide a reasonable measure of shell discard intensity (Bailey 1994; Parkington and Brand 2020), these differences raise the possibility that mollusk exploitation in the inner coasts of the Beagle Channel had a greater emphasis compared to its outer sector.

Another important factor to take into account is the presence of differing chronological patterns along the southern coast of Tierra del Fuego. Radiocarbon dating of shell-bearing sites indicates a correlation between the formation processes of these deposits and the historical dynamics of human settlement in the region. On the northern coast of the Beagle Channel, shell middens can be traced back to as far as 7500 years ago, with many of them exhibiting extensive temporal sequences spanning thousands of years, as seen in sites like Túnel I and Imiwaia I. In contrast, sites in Moat Bay resulting from intensive shell discard date back to the Late Holocene (around 1000 years ago), indicating shorter periods of accumulation, possibly spanning only hundreds of years. Furthermore, the paleodemographic reconstructions suggest relatively stable conditions during most of the occupation period in the area, extending up to about 2000 years ago, as previously outlined in Chap. 2. However, beyond this point, there is a notable increase in the frequency of radiocarbon dates, with a peak in occupations occurring around 500 years ago, followed by a significant decline. This chronological pattern implies fluctuations in the intensity of human occupation and resource utilization within the region, which, in turn, gives rise to different expectations regarding mollusk exploitation strategies. Firstly, it is anticipated that the sizes of target shellfish species found in Late Holocene archaeological assemblages (post-2000 years ago) would be smaller compared to earlier periods. This assumption is rooted in the idea that the higher frequency of occupations during the indicated period likely resulted in greater harvesting intensity and impact on mussel beds. However, it is also plausible that over the last 500 years, as the frequency of occupations decreased, the intensity of mussel exploitation might have lessened.

Previous studies indicate no signs of mollusk depletion throughout the occupation period (Orquera and Piana 2001; Zangrando et al. 2017; Pérez 2020). However, overexploitation is merely one of several observable effects resulting from human-induced impact when examining long-term scales (Jones and Richman 1995; Mannino and Thomas 2001). While mussels are typically considered R-strategists and known for their resilience to environmental variations (Jones and Richman 1995; Pérez 2020), an increasing level of anthropogenic pressure, driven by demographic or environmental factors, may lead to alterations in the size-age structure of intertidal populations in the area (Lasiak 1992; Mannino and Thomas 2001). Regardless of whether social practices operating in short periods (e.g., the regular mobility of groups), enable rapid recovery of the mollusk stock, demographic and environmental variables act as historical conditioning factors for daily activities. Consequently, it is reasonable to anticipate that, even though Middle and Late Holocene assemblages exhibit similar human mobility and collecting strategies, the size, age, and composition profiles of exploited mollusks might display significant long-term distinctions. In the latter scenario, we can argue that demographic or environmental factors played a substantial role in shaping the composition of malacological assemblages.

Considering that the formation of shell deposits in the area is the result of discontinuous processes and variable accumulation rates (Orquera and Piana 1992; Zangrando et al. 2020), it is unlikely that long-term trends in mollusk exploitation are expressed in the archaeological sequences of the deposits.

4.3 Mollusks: Current Natural Populations

The southern coast of Tierra del Fuego is an integral part of the extensive network of channels and fjords comprising the Magellanic sub-Antarctic ecoregion. Within this biogeographic domain, the coastal ecosystem exhibits remarkable environmental diversity, marked by fluctuating salinity and temperature gradients, as well as distinct seasonal patterns (Rosenfeld et al. 2013). Intertidal areas with hard bottoms composed of boulders and rocks predominate, interspersed with sandy beaches or muddy terrain. This environmental heterogeneity encourages a vast biodiversity within benthic assemblages, particularly among mollusk communities (Ojeda et al. 2014). In the rocky shores sessile filter feeders from the Mytilidae family comprise the most abundant mollusc taxa on an annual basis. Gastropods (snails and limpets) and polyplacophores follow them in relative abundance. Unlike their counterparts, mussels maintain relatively stable year-round populations (Langley et al. 1980; Ojeda et al. 2014) and cluster together to form dense aggregations that provide a substrate for various other organisms (Moreno et al. 1984; Castilla and Durán 1985) and facilitate large-scale removal. The bathymetric distribution of mussels encompasses both intertidal and subtidal zones. The former includes mixed banks of *M. edulis* and *B. purpuratus*. At greater depths, a matrix of *M. edulis* and *A. atra* prevails.

In particular, the eurythermic *M. edulis* is the most abundant taxon on the central-east southern coast of Tierra del Fuego (Forcelli 2000; Ojeda et al. 2014). This species has a broad geographic distribution in temperate waters across both the northern and southern hemispheres, exhibiting tolerance to temperatures ranging from 0 to 29 °C (Gosling 1992). In South America, the blue mussel's natural habitat spans the Pacific coast from central Chile, around Cape Horn, and moves northward along the Atlantic coastline until reaching its northern limit in southern Uruguay (McDonald et al. 1991). Within the Beagle Channel, these mussels thrive in conditions characterized by temperatures fluctuating between 3 and 12 °C and salinity levels ranging from 23 to 35‰, as reported by Obelic et al. (1998), Colonese et al. (2012) and Nicastro et al. (2020).

Blue mussels inhabit hard substrates to which they attach themselves using their byssal threads, such as exposed rocky platforms subject to wave action and pebbles or cobblestones (Lasta et al. 1998; Zaixso et al. 2015). It is also possible to find mussel beds on sandy beaches or hardened mud with supporting elements like gravel and empty shells that facilitate their attachment (Gordillo 1991: 129). The ecological plasticity characterizing this species enables it to colonize microhabitats with varying levels of salinity, temperature, and oxygen concentration (Benedetti-Cecchi and Cinelli 1997; Ojeda et al. 2014). *M. edulis*' vertical distribution seems to be influenced by both biological and physical factors. The lower limit is shaped by the presence of predators like crabs and starfish intricately related to the coastal topography, while the upper boundary primarily depends on physical factors such as thermal stress and desiccation (Seed 1969a, b; Seed and Suchanek 1992). Furthermore, mussel colonies located at depths that are challenging for human harvesting act as reserves or meta-populations of significant importance for the replenishment of intertidal beds, especially in the face of anthropogenic impact (Lasta et al. 1998; Orquera and Piana 1999: 117).

M. edulis is a dioecious species with external fertilization that exhibits a prolific reproductive strategy. It releases a substantial quantity of eggs and planktonic larvae with the inherent ability to endure in the water column for months and spread extensively through ocean currents (Seed and Suchanek 1992; Lasta et al. 1998). Like other marine invertebrates, their reproductive cycle hinges on external factors—such as temperature, salinity, photoperiod, and feeding—and internal factors, including nutrient reserves, hormonal cycles, and genome, all of which are subject to geographic variations (Seed 1969a; Tortorelli 1987; Seed and Suchanek 1992). In temperate zones, the impact of seasonality condenses the sexual cycle to a limited span of a few months each year. Within the same population, there's a notable synchronicity in the maturation phases among individuals, and these phases are intricately influenced by temperature gradients (Seed and Suchanek 1992; Lasta et al. 1998). The seasonal peaks of maximum gamete emission into the water column align with the rise in water temperature (typically occurring between 10 and 13 °C) after the winter minimum (Gray et al. 1997; Oehrens Kissner and Kroeck 2005). However, the timing and specific duration of these peaks exhibit variability across different locations (Trancart 1978; Lasta et al. 1998; Kissner and Kroeck 2005). In the Beagle Channel, Tortorelli (1987) observes that spawning takes place from early autumn to early summer, with a pronounced peak between late autumn and early winter, and another, more significant one during spring. On the other hand, Gray et al. (1997) note a singular spawning period during the summer season (December–March) for mussels in the Malvinas Islands mesolittoral. Following this period, individuals enter a phase of sexual rest until the subsequent austral spring (August–September).

Trancart (1978) and Lasta et al. (1998) determined that juvenile individuals from the Atlantic coast of Argentina reach first sexual maturity at one year of age, reaching a size of approximately 30 mm in length. However, it's worth noting that other studies have reported smaller sizes, such as 16 mm (Vinuesa 1979) and 24.78 mm (Oehrens Kissner and Kroeck 2005).

Individual growth rates of *M. edulis* exhibit significant variability, influenced by both ontogenetic age and environmental conditions (Seed 1969b; Seed and Suchanek 1992). Concerning the first factor, as mussels age, somatic growth decreases and/or ceases; conversely, their reproductive activity and potential increase (Seed and Suchanek 1992; Lasta et al. 1998). In the Beagle Channel, mean length values of 21.0 ± 6.2 mm, 57 mm, and 72.9 ± 4.8 mm have been documented 12, 26, and 38 months after the attachment of postlarvae to shell beds (Orquera and Piana 2001: 356). Additionally, preliminary results from an ongoing study on the growth of intertidal mussels in Moat Bay indicate a notable shift between 3 and 4 years of age, marked by increased overlap among different size classes (Fig. 4.2). In neighboring areas of the South American Atlantic coast, other researchers report that specimens of this species achieve a commercial size of 6 cm in length between 2 and 3.5 years of age (Trancart 1978; Bala 1989).

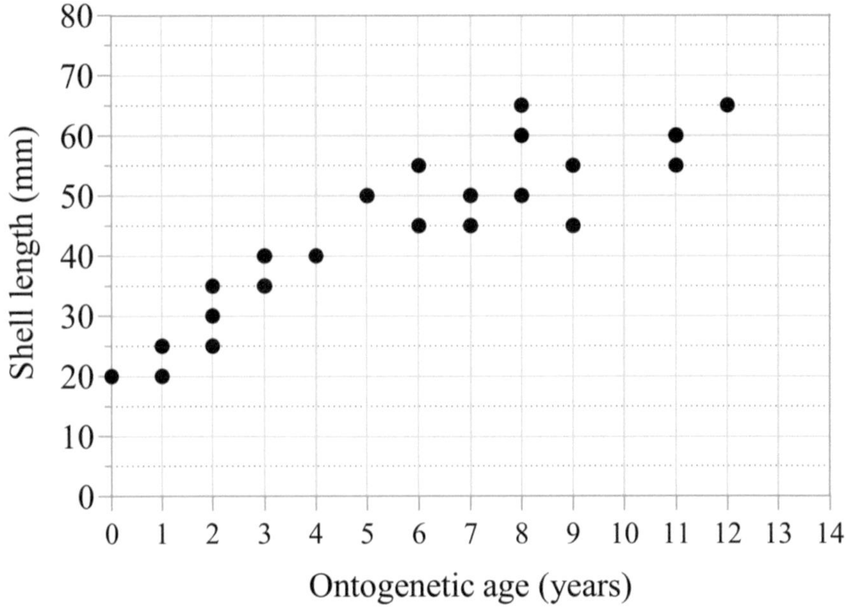

Fig. 4.2 Size-at-age distribution for modern samples of *M. edulis* collected in Moat Bay

The growth and morphology of *M. edulis* are significantly influenced by various ecological conditions in its environment, fostering considerable phenotypic plasticity in the species. Optimal growth for mussels occurs in marine environments with temperatures ranging from 10 to 20 °C and salinity levels between 27 and 35 psu (practical salinity units, Bayne 1976; Wanamaker et al. 2007). Temperature directly affects mussels metabolism, with records outside the optimal range negatively impacting growth, particularly in the embryonic and larval stages, where individuals are more sensitive to environmental variations (Tortorelli 1987; Mestre et al. 2009). However, within the tolerance limits of the species, an increase in seawater temperature contributes to accelerating the growth of somatic and reproductive tissues (Page and Hubbard 1987).

Several ecological studies have observed spatial and temporal variations in growth rates, primarily influenced by the combined effects of water temperature and food abundance (Blanchette and Gaines 2007; Menge et al. 2008). Food availability, in turn, depends on phytoplankton blooms, upwelling cycles, and ocean current circulation patterns (Page and Hubbard 1987; Blanchette and Gaines 2007). Generally, under conditions of higher food concentration, increased water temperature promotes faster growth rates and/or larger sizes in mussels (Menge et al. 2008). This relationship explains the seasonal fluctuations in mussel growth, with acceleration during the productive seasons of spring and summer and a decrease and/or cessation in winter under cold and food-scarce conditions (Seed 1969a, b).

At the spatial level, differences in the shell lengths of individuals from populations in different intertidal sites correlate with spatial gradients in these ecological parameters (Page and Hubbard 1987; Menge et al. 2008). In sub-Antarctic waters, Gray et al. (1997) identified variations in the size-age relationship between colonies of *M. edulis* settled in different sites at the same intertidal level. These variations were associated with local differences in food availability; for instance, in one locality, mussels reached a length of 50 mm at four years of age, while in the other two surveyed sites, the length during the same period was 40 mm.

The variations in depth also contribute to differences in the size attained by blue mussels. Individuals growing in the intertidal fringe experience harsh environmental conditions that impact their metabolic rates, resulting in slower growth, delayed maturation, and smaller sizes compared to subtidal counterparts in the infra- and circumlittoral zones. Concurrently, their reproductive contribution is significantly lower than that of subtidal individuals, as they invest a smaller percentage of body tissue in gametes. These size differences are also evident within different intertidal levels. Sampling across various sectors of the southern coast of Tierra del Fuego reveals smaller average lengths in the high intertidal strip compared to the middle and lower intertidal levels (Table 4.1). Furthermore, there are variations in allometric relationships based on depth. Along the coasts of the Beagle Channel, mussels from the intertidal zone develop more robust and rounded shells than those suspended at greater depths, characterized by shells that are higher, lighter, and narrower (Silva and Calvo 1995).

In terms of the population dynamics of *M. edulis*, individuals of this species exhibit a spatially structured distribution along the coast. Mussel beds consist of

Table 4.1 Descriptive statistical parameters for contemporary intertidal mussel (Mytilidae) shell sizes across various sampling locations along the south coast of Tierra del Fuego

Locality	Intertidal zone	*n*	Mean (mm)	SD	Min (mm)	Max (mm)	Source
Túnel-L. Packewaia	Middle	5401	23.20	9.80	4.00	66.00	Orquera and Piana (2001)
	Low	827	36.60	16.10	7.00	68.00	
Remolino	Middle	163	54.70	6.00	31.00	68.00	
	Low	718	48.20	10.80	15.00	79.00	
Imiwaia	Middle	814	23.60	6.30	9.00	45.00	
	Low	579	48.80	8.50	12.00	71.00	
Inner Cambaceres	Middle	2012	22.70	9.50	3.00	48.00	
	Low	448	42.20	16.10	10.50	79.00	
Outer Cambaceres	Middle	2437	15.20	9.10	4.00	63.00	
	Low	1497	31.80	21.10	4.00	75.00	
Heshkaia	High	269	35.90	5.50	17.90	53.60	Zangrando et al. (2017)
	Middle	122	44.80	5.40	31.30	61.40	
	Low	62	59.60	6.00	44.40	70.60	

individuals from various recruitments, forming discrete and segregated groupings that are, nevertheless, perpetually interconnected through planktonic larvae. The demographic structure of these beds can vary significantly spatially, both among different coastal localities and within the same habitat, owing to differences in recruitment cycles, growth rates, and mortality rates of individuals (Seed 1969a, b). In subantarctic waters, Gray et al. (1997) observe poly-, bi-, and unimodal patterns in the length-frequency distributions of populations settled near similar microenvironmental conditions. The settlement of postlarvae onto the substrate occurs throughout the year, primarily between December and February (Bala 1989; Orquera and Piana 2001: 356). The relatively lengthy recruitment periods and variability in individual growth rates lead to an overlap in cohort size, rendering cohorts indistinguishable (Seed 1969a, b; Gray et al. 1997). These aspects caution against relying solely on size frequency distribution curves for characterizing population structure due to the resulting limitations.

In summary, various ecological studies suggest that *M. edulis* is a sensitive but highly resilient species to environmental variation (Seed 1969a, b; Suchanek 1981; Seed and Suchanek 1992). The biology and life history of blue mussels, comparable to those of other species in the genus Mytilidae, categorize them as R-strategy resources. In this regard, they are characterized by short life cycles, high reproductive rates, and rapid growth.

4.4 Mussel Archaeological Assemblages and the Interpretation of Shell Sizes

Archaeofaunal shellfish research focusing on the importance of these resources in past human subsistence is usually built upon metrical analysis (Claassen 1998; Faulkner 2013). In particular, the shell length dimension is typically used as a proxy for the body size and meat content of targeted individuals (Anderson 1981; Erlandson 1991; Klein et al. 2004; Codding et al. 2014). In southern Tierra del Fuego, this data was initially obtained through in situ measurements of whole specimens recovered from 40 × 20 cm sampling columns excavated from shell middens located in the Beagle Channel (Fig. 4.3; Orquera and Piana 1992, 2000). Mussel shell length was systematically measured at regular intervals, each 5 cm in depth, and using manual calipers. This method circumvented potential biases in mean length estimations arising from measuring only whole shells, a concern due to the fragmentation caused by the removal and transport of the samples to the laboratory (Orquera and Piana 2000). While the measurement task included mussels from different Mytilid taxa, such as *M. edulis*, *B. purpuratus*, and *Aulacomya atra*, the dietary significance of these bivalves was essentially discussed at the family level by calculating global length averages derived from all recorded sizes, and without taxonomic discrimination by species.

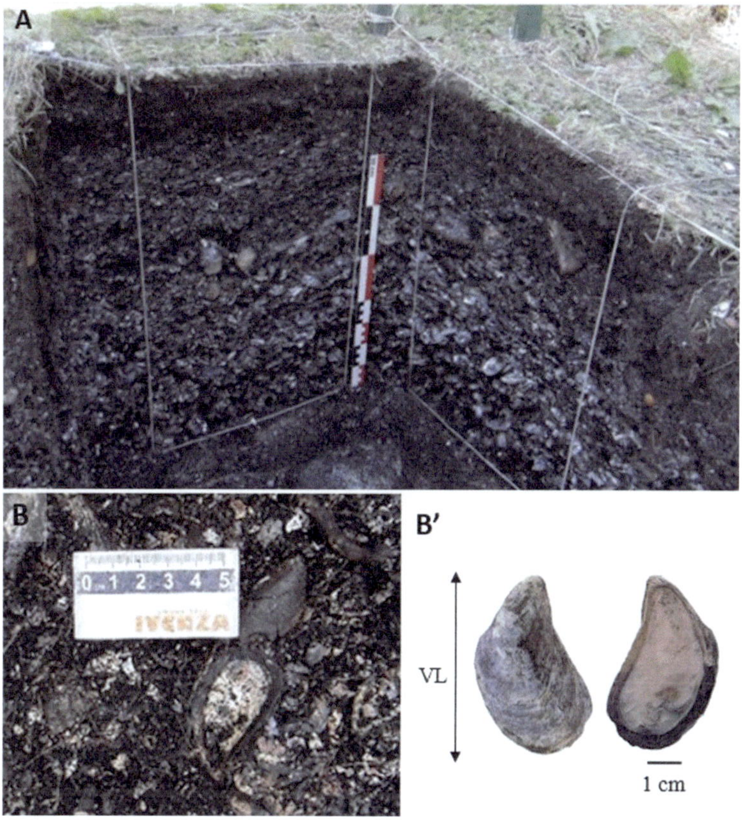

Fig. 4.3 Column for in-situ measurement of mussels in a shell midden from the study area (**a**); an archaeological *M. edulis* specimen (**b**) and its total valve length (VL) dimension (**b'**)

The nutritional contribution of archaeological mussels from the Beagle Channel region was assessed based on data obtained from living *M. edulis* individuals collected in the intertidal and slightly subtidal zones near the archaeological sites (refer to Table 4.1). According to this actualistic information, the optimal sizes that can be gathered by using procurement techniques similar to those described in ethnographic records are about 57.3 ± 6.3 mm long (Orquera and Piana 2001: 355). However, measurements obtained from various shell middens in Túnel, Shamakush, and Imiwaia revealed a prevailing presence of smaller mussels with average length values ranging from 32.0 ± 7.7 to 49.3 ± 10.8 mm (Orquera and Piana 2001: 355). These sizes suggest that mussels were mostly collected during the second year after their attachment to the substrate (Orquera and Piana 2001: 356). Moreover, because valves larger than 80 mm are rarely documented in shell middens, it is assumed that mussel harvesting predominantly occurred in shallow waters, reaching depths of up to two meters, where mussels exhibit slower individual growth rates than at lower shore levels (Orquera and Piana 1999: 117).

When examining modern *M. edulis* through biochemical analysis, it becomes evident that individuals with sizes resembling those found in archaeological samples offer a reduced caloric yield compared to those attaining optimal sizes. The latter provide, on average, 2.67 kcal and 0.45 g of protein per individual, whereas mussels around 43 mm yield 1.45 kcal per individual and 0.27 g of protein (Orquera 1999: 311; Orquera and Piana 2009). Therefore, based on this information, previous studies suggest that shellfish was suboptimally exploited by human groups during the second half of the Holocene (Orquera 1999; Orquera and Piana 2001). In addition, mussels measuring below 30 mm exhibit very low meat content (less than 0.66 g of cooked meat and 0.26 g of dehydrated meat), rendering them practically valueless economically. Their presence in shell midden deposits has been interpreted as incidental, likely resulting from the extraction of entire sections of the seabed, which removed individuals indiscriminately, regardless of size (Orquera and Piana 1993–94; 2001).

To account for the diminished size of archaeological mussels compared to those presently found in the intertidal zone, previous research in the Beagle Channel has delved into the influence of environmental factors, including the thermal conditions of surface seawater (Orquera and Piana 2001). This is noteworthy since sea surface temperature has important effects on the structure of benthic communities and directly influences crucial biological functions in marine organisms, including growth (Blanchette et al. 2006). In this context, the isotopic study of *Mytilus edulis* samples from shell midden sites in the Beagle Channel offers a comprehensive record of the thermal fluctuations of surface seawater over the past six thousand years (Obelic et al. 1998). Between 6800 and 5700 cal yr BP, the paleotemperature curve indicates cooler conditions compared to the present, with a thermal record 1.5 °C below the current annual average. Subsequently, a temperature increase is noted, reaching a maximum between 5100 and 4400 cal yr BP, with slightly higher values than today. This period aligns with the global climatic amelioration known as the Holocene Thermal Maximum or Hipsithermal. Following this, there is a cooling of the waters peaking around 3700 cal yr BP, with a temperature 1 °C lower than the current average, followed by another warm phase between 3100 and 2200 cal yr BP. From this point onward, there are several intervals of cooler conditions between 1900 and 800 cal yr BP, and between 500 and 100 cal yr BP, along with warm conditions between 800 and 500 cal yr BP. These last two periods align respectively with the Little Ice Age and the Medieval Climate Anomaly, climatic events identified at other latitudes (Obelic et al., 1998).

Isotopic data indicates that in the last 6800 cal yr BP, the Sea Surface Temperature (SST) of the Beagle Channel fluctuated between 1 and 0.5 °C relative to the current annual mean (6.5 °C). With this information, it has been argued that these variations would not have significantly impacted mussel growth patterns, as the magnitude of the changes closely aligns with the current range of seasonal thermal fluctuations (Orquera and Piana 2001). While there is not enough information to determine the extent to which fluctuations in seawater temperature may have affected mussel growth rates, the identified paleoenvironmental changes imply different conditions for the development of these mollusks and may not sufficiently explain the consistently smaller size of specimens observed throughout the regional archaeological sequence.

Instead, the primary cause for archaeological shell dimensions below those found in natural populations not exposed to human predation has been attributed to intense human harvesting pressure. Modal sizes documented in the archaeological record suggest frequent massive collections in intertidal mussel beds as soon as mussels reached a minimum profitable size (Orquera and Piana 2001; 2009). Basing on some size-age observations (Silva 1996) and nutritional information (Orquera 1999), this means that for these mollusks, within a year or less, the remaining uncollected small specimens would have grown enough to reach an adequate size for human consumption (Orquera and Piana 2009: 77). By lowering the numbers of large specimens, the intensive mussel exploitation would have resulted in reduced mean shell size and, probably, average age among the specimens remaining within the intertidal patches (Orquera and Piana 2009: 77). In addition, regardless of whether mussels were collected by size selection or by stripping, sustained, high levels of exploitation limit the opportunity for mollusks to grow larger, contributing to this effect of reduced age and size changes in shell beds. As a consequence, the archaeological prey population reflects the decreasing mean shell size of the natural population being exploited (Mannino and Thomas 2001; Giovas et al. 2013).

Because mussels prevail in all shell middens along the coast of Beagle Channel, it is evident that local reproduction and population levels remained sustainable despite the depleting effects of human predation. Several reasons have been invoked for this. Regular larval recruitment from healthy populations living in subtidal zones and outside local waters would have helped repopulate discrete patches under human impact and maintain stable population levels (Orquera and Piana 1999; Orquera 2005). In addition, shell midden formational processes point to a short-term site occupation pattern and high mobility implying that intensive exploitation of local resources was temporal (Orquera and Piana 1989–90, 2009; Orquera 1999). Recent studies on oxygen isotopic composition ($\delta^{18}O$) of archeological limpet shells have also suggested intertidal shellfish in the Beagle Channel were exploited periodically and not year-round (Colonese et al. 2011, 2012). In this context, some temporal gaps would have allowed local mussel colonies to recover and stabilize, thereby ensuring their long-term productivity. Finally, estimated mean sizes in archaeological samples suggest that mussels would have reached sexual maturity before being harvested so the exploited shore would always have had reproducing individuals (Lasiak 1991), which contributed to the survival of the local population. In sum, shellfish depletion was a localized process with no ecological effects at larger-scales such as the metapopulation level.

Subsequent investigations conducted on shell assemblages with post-contact chronologies (spanning from the late eighteenth to the nineteenth centuries) have enhanced our understanding of the anthropic pressure imposed on malacological resources toward the conclusion of the occupational sequence in the central Beagle Channel (Verdún Castelló 2011, 2014; Colonese et al. 2012). The average lengths of complete mussels recovered from extensive excavations at the Late Holocene sites of Túnel VII and Lanashuaia (31.5 and 35.3 mm, respectively), were found to be approximately 10 to 20 mm smaller (Verdún Castelló 2014: 234).

Further east in Moat Bay, shell dimensions from archaeological deposits dating between 900 and 200 cal yr BP also indicate suboptimal and intensive exploitation of mollusks (Zangrando et al. 2017; Pérez 2020). Previous studies conducted at the Heshkaia locality estimated valve total length from fragments recovered in extensive excavations by applying allometric formulas to umbos dimensions proposed by Campbell (2012). Results indicated average shell length values between 31 and 38 mm (Table 4.2; Zangrando et al. 2017; Pérez 2020). Similar to what was observed in the Beagle Channel, archaeological specimens in Heshkaia shell middens are smaller than those recorded in the surrounding intertidal beds. According to the local ecological information, during low tides, it is possible to collect individuals with maximum sizes up to 70 mm within the lower intertidal, where the average length is 59.6 ± 5.9 mm (Zangrando et al. 2017). The means and size ranges of shells at these sites closely resemble those recorded in other late archaeological shell middens from the central Beagle Channel, suggesting no differential patterns between these geographic sectors in terms of the size of targeted mussels.

4.5 Shell Size Variation in the Long-Term Scale

The size variability observed in the archaeological shell accumulations along the southern coast of Tierra del Fuego has been primarily discussed from a diachronic perspective. In the Beagle Channel region, previous studies have noted fluctuations in mussel size both throughout the individual stratigraphies of sites and between sites, with no indication of a clear and consistent long-term trend (Orquera and Piana 1999, 2001). In multi-component sites with chronological sequences spanning thousands of years (e.g., Túnel I, Imiwaia I), the largest mean lengths are represented at different depths of the deposits (Orquera and Piana 1999, 2001: 357–359). This contrasts with typical scenarios of anthropogenic overexploitation, characterized by the progressive reduction of modal sizes from basal to upper stratigraphic levels of the sites (Mannino and Thomas 2002). Likewise, in deposits accumulated over shorter periods (e.g., Shamakush X, Shamakush I, and Lancha Packewaia I), no discernible trends suggesting the gradual depletion of shellfish resources are apparent.

Previous studies have also highlighted variations in mussel size across different locations in this region. The observed trends reveal distinct local scenarios of anthropogenic impact on intertidal mussel populations. These observations are based on the statistical comparison of size frequency distributions of chronologically discrete deposits instead of the sites as a whole, and in all cases avoiding normality assumptions of data. In both Túnel-L.Packewaia and Imiwaia localities, the maximum sizes identified in the early stages of the occupational sequence are absent in later deposits, indicating an apparent overall reduction in valve length (Table 4.2). These findings suggest a gradual increase in anthropogenic pressure on local mussel beds over time. In Túnel-Lancha Packewaia, statistical analysis confirms a progressive reduction in sizes, with significant differences among the median lengths of all archaeological assemblages with dissimilar chronologies (Kruskal–Wallis, $X2 = 824.4$,

Table 4.2 Age and descriptive statistics for mussel (Mytilidae) shell assemblages from different archaeological localities along the south coast of Tierra del Fuego reported by Orquera and Piana (2001); Verdún Castelló (2014); Zangrando et al. (2017); and Pérez (2020)

Locality	Archaeological shell assemblage	Calibrated years BP (Median probability)	*n*	Mean (mm)	SD	Median (mm)	Modal range (mm)		Total range (mm)	
							min	max	min	max
Túnel-L. Packewaia	Túnel I (D: XIV)	6790	2343	41.54	12.18	42.00	35.00	49.00	7.00	94.00
	Túnel I (D: IV-XVI)	6575–5290	1713	38.51	10.40	40.00	31.00	45.00	7.00	90.00
	L. Packewaia (X)	4450	254	43.80	9.97	43.50	38.00	49.25	12.00	78.00
	L. Packewaia (D)	940	1535	35.65	10.55	35.00	28.00	43.00	6.00	78.00
	Túnel II	990	169	31.98	7.68	32.00	28.00	37.00	9.00	56.00
	L. Packewaia (B)	290	488	39.75	9.30	40.00	32.00	46.00	15.00	66.00
	Túnel VII	Modern	1764	33.39	9.73	32.00	27.00	40.00	5.00	74.00
Imiwaia	Imiwaia I (K, L, M)	6730	490	49.33	10.88	48.00	43.00	55.00	12.00	104.00
	Imiwaia I (D)	3710	46	56.22	9.33	56.00	49.50	63.25	37.00	73.00
	Imiwaia I (B)	1340	95	45.87	9.08	45.00	40.00	49.00	33.00	76.00
Lanashuaia	Lanashuaia I	1030	158	33.49	9.8	33.19	29.00	40.01	3.96	60.75
Shamakush	Shamakush I	820	1928	40.21	10.54	39.00	33.00	46.00	17.00	82.00
	Shamakush X	485	2374	47.77	10.01	48.00	42.00	54.00	6.00	79.00
Heshkaia	Heshkaia 34	850	83	38.60	5.35	38.05	34.29	41.69	28.62	53.93
	Heshkaia 35	700	408	31.65	7.14	31.72	27.36	35.92	13.31	60.23
	Heshkaia 30	215	259	36.73	6.13	36.70	32.70	41.00	23.00	54.20

$p < 0.001$), except for Túnel II and Túnel VII sites, which exhibit similar values (Mann–Whitney U, $p = 0.158$). In the Imiwaia locality, mussel dimensions from shell middens are larger than those recorded in Túnel-L. Packewaia (Table 4.2). However, the most recent shell midden also displays markedly smaller values compared to Mid-Holocene levels (Mann–Whitney U: $Z = 3.8$, $p < 0.001$), although these changes are not expressed gradually. Lastly, in Shamakush, available metric information is so far restricted to two Late Holocene archaeological sites, the outlined trend diverges from previous cases, with significantly larger specimens represented in the most recent assemblage (Mann–Whitney: $Z = 24.8$, $p < 0.001$).

The presence of large/old mussels in the late contexts of Imiwaia and Shamakush, with dimensions comparable to Mid-Holocene contexts, suggests that mussels could still recruit at older ages in many parts of the region, and that the life-history of these bivalve species would have been virtually similar to earlier times. Coupled with the absence of temporal decline in mussel relative abundance, these observations suggest the metapopulation was ecologically stable over time, contravening the idea of a diachronic regional process of shellfish overexploitation (Orquera and Piana 1999; 2001).

In the eastern sector of the Beagle Channel, archaeological deposits from the Heshkaia locality (Moat Bay) also exhibit an oscillatory pattern in the modal and in the average size of mussel individuals within the stratigraphic sequences (see Table 4.2). In particular, the Heshkaia 35 site also exhibits decreasing median values toward the upper levels of the deposit (Zangrando et al. 2017). In this latter site, the overall average length is lower than that identified in sites with earlier dates (Heshkaia 34) and later ones (Heshkaia 30), suggesting a stable local utilization of these resources over time (Pérez 2020).

At a macro-regional scale, a great variability is observed in the dimensions of valves across all the examined shell middens along the south coast of Tierra del Fuego. In these deposits, size distributions exhibit significantly different median values (Kruskal–Wallis test: $X2 = 2984$; $p = 0$). However, diachronic variation in the modal size of mussels displays a highly irregular pattern throughout the archaeological sequence of the area (Fig. 4.4). Additionally, grouping the length measurements of valves into two temporal blocks indicates that the median value of Late Holocene mussels (3700–100 cal yr BP) is significantly lower than that of specimens from the Mid-Holocene (6800–4200 cal yr BP) (Mann–Whitney test: $U = 1.98E + 07$, $z = 10.903$, $p < 0.001$). In this latter period, there is a greater dispersion of lengths, and higher maximum values in overall size ranges are recorded. Shells over 80 mm in length may be indicative of *A. atra* specimens, as this species, predominantly subtidal in habit, tends to exhibit larger dimensions compared to other mussel taxa. These measurements are primarily identified in early deposits and also coincide with comparatively higher abundance frequencies for these bivalves compared to later periods. Nevertheless, statistical analysis of Spearman's rho indicates a weak and non-significant correlation between the abundance indices of *A. atra* and the overall mean lengths of mussels (including all taxa) calculated across all the archaeological assemblages (Fig. 4.5; $rs = 0.032$, $p = 0.920$). Similarly, the correlation between mean sizes and the abundance indices of small taxa like *B. purpuratus* also lacks statistical significance (rs = – 0.507, $p = 0.092$). The findings imply that the presence of secondary mussel taxa has minimal influence on the identified size patterns in shell assemblages. Conversely, the robust correlation between valve lengths and the abundance of small specimens of *M. edulis* (refer to Fig. 4.5; rs = – 0.82, $p < 0.001$) suggests that the long-term variation in valve dimensions is predominantly influenced by the harvested sizes of this species.

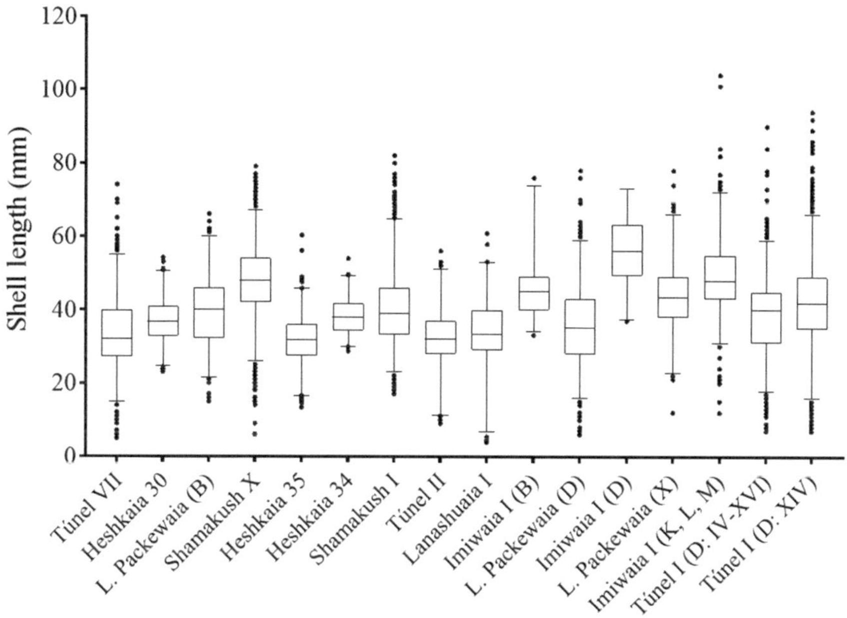

Fig. 4.4 Boxplots showing the distribution of the variable length for mussels from different archaeological assemblages arranged by chronology from left (Late Holocene) to right (Middle Holocene)

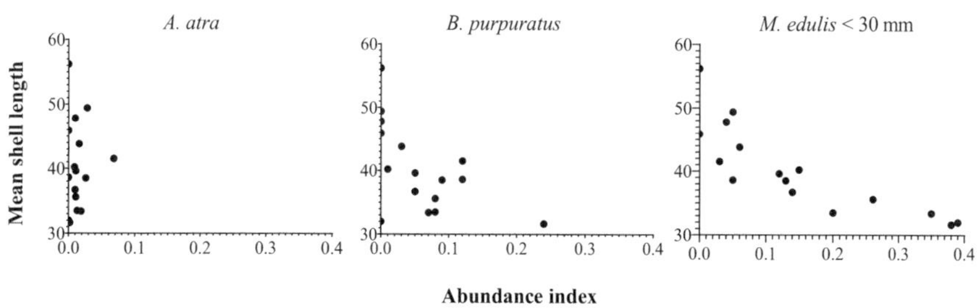

Fig. 4.5 Mussel mean valve size relative to abundance indexes for distinct mussel taxa, each point corresponding to temporally discrete archaeological assemblages from the study area in Tierra del Fuego

Previous studies aimed to explain diachronic fluctuations in the archaeological size of mussels by analyzing the degree of correlation between sizes in shell middens and the paleoclimatic fluctuations during the second half of the Holocene in the area (Zangrando et al. 2017; Pérez 2020). In coastal areas affected by upwelling processes, cold climatic periods are often construed as scenarios of enhanced coastal productivity, conducive to increased density and growth rates of mollusc populations (Erlandson et al. 2011; Flores et al. 2018). These assumptions are grounded in ecological information indicating, on the one hand, a positive correlation between the influence of nutrient-rich deep waters and marine primary productivity (Blanchette et al. 2006), and on the other hand, an inverse correlation between ambient temperature and metabolic rates of various malacological species (Seed 1968; Griffiths 1981; Kilburn and Rippey 1982). However, in the Beagle Channel, paleoecological studies have revealed that the longevity, growth rates, and sizes of fossil shells of some benthic bivalve species were higher during the warm period of the Climatic Optimum (4400 cal yr BP) compared to those recorded in previous cold periods (5700 cal yr BP) and later (Little Ice Age, Gordillo et al. 2015), as well as those observed today (Morán et al. 2021). The optimal development of these molluscs during a warm period has been attributed to the estuarine dynamics that characterize the waters of the Beagle Channel and the fjord system in general (Gordillo et al. 2015; Morán et al. 2021). In this geographical area, coastal primary productivity depends on both the advection of saline oceanic waters from the Antarctic Circumpolar Current (ACC) and the input of organic-rich meltwater from terrestrial sources (Iachetti et al. 2021). Therefore, it is inferred that during periods of climate warming, increased discharge of continental waters would have promoted higher primary productivity in the Beagle Channel and the coastal environment in general (Gordillo et al. 2015; Morán et al. 2021).

However, the size variability observed in the shells of archaeological deposits during different climatic periods is wide and does not consistently reflect the thermal fluctuations of oceanic water over the last 6800 cal yr BP (Fig. 4.6). In this regard, larger sizes, above 40 mm, are represented under very different SST values. Climatic fluctuations may have exerted a diverse impact on the productivity of mollusk beds, due to the local dynamics of oceanographic conditions, as observed in other research areas (Flores et al. 2018). Nevertheless, even with these potential variations, local archaeological sequences do not distinctly align with size fluctuations in relation to the alternating cold and warm climatic intervals. Essentially, this information indicates that changes in SST do not provide a comprehensive explanation for the observed long-term size changes in the archaeological record along the south coast of Tierra del Fuego. Although environmental changes might have had a role on mussels' growth rates, they cannot alone explain the size patterns archaeologically observed.

In contexts where paleoenvironmental factors can be dismissed, numerous studies have focused on discussing size patterns observed in archaeological shell accumulations in the light of anthropogenic processes related to broader economic and demographic changes (Mannino and Thomas 2002; Erlandson et al. 2011; Faulkner 2013). In relation to this possibility, it is worth noting that for the south coast of Tierra del Fuego, zooarchaeological and paleodemographic evidence reveals more intensive

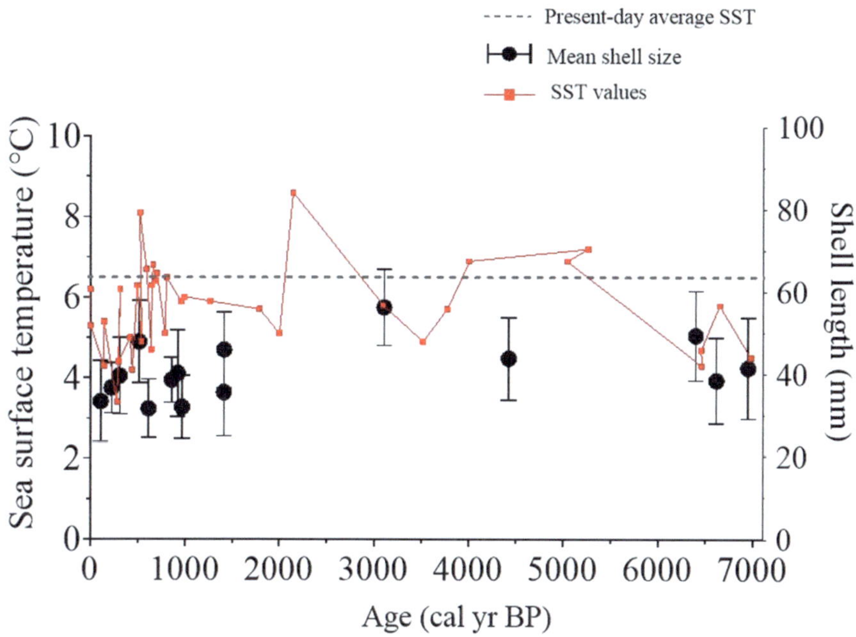

Fig. 4.6 Paleo water temperature (left axis) estimated by Obelic et al. (1998) and mean shell size values (right axis) of *M. edulis* specimens of different archaeological assemblages

use of spaces and new modes of provisioning certain coastal resources (e.g., fish and birds) from around 2000 cal yr BP (Zangrando 2009; Zangrando et al. 2020). In the realm of malacological fauna, the diminished sizes evident in the modal lengths of mussels within deposits such as Túnel II and VII, Lanashuaia I, and Heshkaia 30 and 35 denote a decline in overall return rates from harvesting practices after 1500 cal yr BP. This decline might be attributed to intensified predation, involving a greater inclusion of less profitable prey (smaller/younger individuals) into the diet compared to earlier times. The smaller sizes could also suggest that more frequent or intense human harvesting prevented mussels from reaching the older ages observed in earlier times. This situation aligns well with the broader context of economic intensification suggested by other lines of evidence. Another noteworthy aspect is the divergence in median sizes among all early assemblages of Túnel I (layer D), Imiwaia I (layers K, L, M, and D), and Lancha Packewaia (layer X), as evidenced by statistically significant differences in Mann-Whitney pairwise comparisons ($p < 0.05$; Table 4.3). In contrast, Late Holocene assemblages exhibit a higher degree of similarity in shell size distributions, showing no statistically significant differences ($p > 0.05$; Table 4.3) in modal sizes among shell middens from across various coastal locations in the area. These observed similarities find resonance in a hypothetical scenario of increased levels of harvesting intensity, with human predation having a more pronounced impact on the age and size composition of intertidal mussel beds.

Conversely, the significant variation among all Mid-Holocene shell assemblages may reflect the inherent natural diversity among shellfish colonies in different intertidal habitats, which are less impacted by anthropogenic exploitation.

Another metric frequently used in archaeological research aimed at investigating temporal changes in the intensity of human exploitation of mollusks is the relative representation of small, reproductively immature specimens compared to adults (larger specimens). The proportional rise in small specimens with limited edible content not only signifies reduced returns in collecting activities but also suggests a more pronounced ecological impact on natural mollusk populations. The inclusion of reproductively immature specimens in mollusk harvesting has the potential to diminish the reproductive capacity and recruitment potential of the species (Lasiak 1991; Mannino and Thomas 2002). In the case of *M. edulis*, indications point to the fact that individuals providing a minimally viable food yield for human consumption are typically around 30 mm in size (Orquera 1999). This measurement corresponds to the maximum size reported for the first sexual maturity of juveniles (Lasta et al. 1998) and can, therefore, be practically employed as an indicator of the ecological impact of human collections. However, even though archaeological findings suggest a decrease in the modal size of collected mussels in later occupation periods, these shifts don't seem to correlate with an increase in the abundance of specimens with dimensions below 30 mm, and presumably juveniles. On the contrary, these smaller individuals appear to be present in variable proportions throughout the macro-regional occupational sequence (Fig. 4.7a).

The possibility that variations in the relative proportion of mussels with limited edible content are attributed to preservation biases seems improbable for several reasons. Firstly, these differences persist regardless of the chronology, depth, and matrix characteristics of various shell middens. Moreover, even though adult specimens typically exhibit greater thickness and resistance than juveniles (Zuschin et al. 2003), certain prior studies have failed to identify significant differences in relative susceptibility to breakage between size/growth-stage categories (Jerardino and Navarro 2008; Cannon and Burchell 2009; Jerardino 2018). These studies have argued that taphonomic processes, such as fragmentation, might introduce a bias toward juvenile specimens, as their small dimensions could make them less susceptible than larger individuals to external force pressures. Finally, it is noteworthy that both in situ valve measurements and the application of regression equations from mussel fragments are strategies designed to minimize biases in size calculations resulting from sample removal, transportation to the laboratory, and/or post-depositional processes (Orquera and Piana 2000; Campbell 2012).

Different locations along the southern coast of Tierra del Fuego simultaneously depict variable situations regarding the abundance indices of these specimens (Fig. 4.7b). In Shamakush and Imiwaia, these specimens exhibit low values, irrespective of the chronological context of the sites, whereas in Túnel-Lancha Packewaia and Heshkaia, there is instead a notable variability in their abundance at the inter-site level. Furthermore, the fact that the modal valve sizes documented throughout the occupational sequence in the area exceed 30 mm in length reinforces the notion that

Table 4.3 Post-hoc Mann–Whitney *U*-tests comparing Mytilidae valve size between different assemblages from Beagle Channel and Bahía Moat regions for Middle (MH) and Late (LH) Holocene

Period	Site pair	*U*	*p-value*
Middle Holocene	Túnel I (D: XIV)-Túnel I (D: IV-XVI)	1694671.5	**< 0.001**
	Túnel I (D: XIV)-Imiwaia I (K, L y M)	354235	**< 0.001**
	Túnel I (D: XIV)-L. Packewaia (X)	262867.5	**0.002**
	Túnel I (D: IV-XVI)-L. Packewaia (X)	153384	**< 0.001**
	Túnel I (D: IV-XVI)-Imiwaia I (K, L y M)	188433.5	**< 0.001**
	Imiwaia I (K, L y M)-L. Packewaia (X)	43012.5	**< 0.001**
Late Holocene	Imiwaia I (D)-L. Packewaia (D)	5377	**< 0.001**
	Imiwaia I (D)-Imiwaia I (B)	884.5	**< 0.001**
	Imiwaia I (D)-Túnel II	155	**< 0.001**
	Imiwaia I (D)-Lanashuaia I	299	**< 0.001**
	Imiwaia I (D)-Shamakush I	11323	**< 0.001**
	Imiwaia I (D)-Heshkaia 34	204	**< 0.001**
	Imiwaia I (D)-Heshkaia 35	312	**< 0.001**
	Imiwaia I (D)-Shamakush X	29501.5	**< 0.001**
	Imiwaia I (D)-L. Packewaia (B)	2477.5	**< 0.001**
	Imiwaia I (D)-Heshkaia 30	467	**< 0.001**
	Imiwaia I (D)-Túnel VII	3844	**< 0.001**
	L. Packewaia (D)- Imiwaia I (B)	32950.5	**< 0.001**
	L. Packewaia (D)-Túnel II	104508	**< 0.001**
	L. Packewaia (D)-Lanashuaia I	112223	0.122
	L. Packewaia (D)-Shamakush I	1133002.5	**< 0.001**
	L. Packewaia (D)-Heshkaia 34	49885	**0.001**
	L. Packewaia (D)-Heshkaia 35	243627.5	**< 0.001**
	L. Packewaia (D)-Shamakush X	7146	**< 0.001**
	L. Packewaia (D)-L. Packewaia (B)	282598.5	**< 0.001**
	L. Packewaia (D)-Heshkaia 30	178723.5	**0.009**
	L. Packewaia (D)-Heshkaia VII	1190969	**< 0.001**
	Imiwaia I (B)-Túnel II	1453.5	**< 0.001**
	Imiwaia I (B)- Lanashuaia I	2381.5	**< 0.001**
	Imiwaia I (B)-Shamakush I	59729.5	**< 0.001**
	Imiwaia I (B)-Heshkaia 34	1898	**< 0.001**
	Imiwaia I (B)-Heshkaia 35	3226.5	**< 0.001**
	Imiwaia I (B)-Shamakush X	91364.5	**0.002**
	Imiwaia I (B)-L. Packewaia (B)	15297.5	**< 0.001**
	Imiwaia I (B)-Heshkaia 30	4785	**< 0.001**

(continued)

Table 4.3 (continued)

Period	Site pair	*U*	*p-value*
	Imiwaia I (B)-Heshkaia VII	26255.5	**< 0.001**
	Túnel II-Lanashuaia I	11418.5	**0.024**
	Túnel II-Shamakush I	86605.5	**< 0.001**
	Túnel II-Heshkaia 34	3202	**< 0.001**
	Túnel II-Heshkaia 35	33378	0.547
	Túnel II-Shamakush X	40256.5	**< 0.001**
	Túnel II-L. Packewaia (B)	21032	**< 0.001**
	Túnel II-Heshkaia 30	13321	**< 0.001**
	Túnel II-Túnel VII	139291	0.158
	Lanashuaia I-Shamakush I	101956	**< 0.001**
	Lanashuaia I-Heshkaia 34	4137.5	**< 0.001**
	Lanashuaia I-Heshkaia 35	26847.5	**0.002**
	Lanashuaia I-Shamakush X	53408	**< 0.001**
	Lanashuaia I-L. Packewaia (B)	25118.5	**< 0.001**
	Lanashuaia I-Heshkaia 30	16195	**0.000**
	Lanashuaia I-Túnel VII	130443.5	0.182
	Shamakush I-Heshkaia 34	75466	0.380
	Shamakush I-Heshkaia 35	199289	**< 0.001**
	Shamakush I-Shamakush X	1283565.5	**< 0.001**
	Shamakush I-L. Packewaia (B)	465416.5	0.7155
	Shamakush I-Heshkaia 30	205620.5	**< 0.001**
	Shamakush I-Túnel VII	1072216	**< 0.001**
	Heshkaia 34-Heshkaia 35	7211.5	**< 0.001**
	Heshkaia 34-Shamakush X	38633	**< 0.001**
	Heshkaia 34-L. Packewaia (B)	18341	0.1689
	Heshkaia 34-Heshkaia 30	8988.5	**0.025**
	Heshkaia 34-Túnel VII	44075	**< 0.001**
	Heshkaia 35-Shamakush X	89728.5	**< 0.001**
	Heshkaia 35- L. Packewaia (B)	48622	**< 0.001**
	Heshkaia 35- Heshkaia 30	30803	**< 0.001**
	Heshkaia 35-Túnel VII	326512.5	**0.003**
	Shamakush X-L. Packewaia (B)	309209	**< 0.001**
	Shamakush X-Heshkaia 30	101523.5	**< 0.001**
	Shamakush X-Túnel VII	610664.5	**< 0.000**
	L. Packewaia (B)-Heshkaia 30	49608	**< 0.001**

(continued)

Table 4.3 (continued)

Period	Site pair	*U*	*p-value*
	L. Packewaia (B)-Túnel VII	264929.5	**< 0.001**
	Heshkaia 30-Túnel VII	166163	**< 0.001**

Bold values indicate statistically significant differences at the 0.05 level

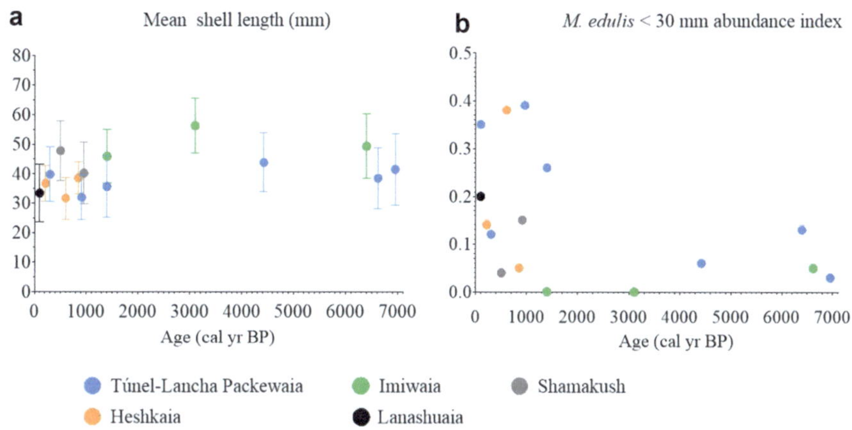

Fig. 4.7 Trans-Holocene changes in mean mussel length measurements (**a**) and abundance of small/non-profitable (< 30 mm) mussels (**b**) shown for shell middens from different archaeological localities

these small-sized mussels were unintentionally collected (Orquera and Piana 2001: 353).

While some data is suggestive, the metric composition of shell accumulations in the area does not provide conclusive evidence regarding the existence of diachronic changes in the degree of exploitation of malacological fauna that can be definitively linked to the history of coastal occupation in the area. This raises the question of what factors underlie and explain the wide variability in the representation of mussel sizes observed in different archaeological assemblages. Prior studies have suggested variations in the frequency of human visits to intertidal banks and natural recruitment cycles as potential causal mechanisms for the permanent oscillation in the average length of collected specimens (Orquera and Piana 2001: 359). Recent exploration in other study areas has delved into the influence of spatial differences in shellfish growth rates on the variability of sizes represented archaeologically (Thakar et al. 2017; Hausmann et al. 2021). These differences are primarily regulated by variations in micro-scale ecological factors, such as intertidal levels, geomorphology, substrate types, freshwater bodies, among others. These studies propose considering the possibility that many temporal trends in valve dimensions are associated with changes in collection zones or patches (for example, obtaining mollusks in different local habitats) rather than the effects of anthropogenic pressure.

Aligned with these interpretations, specific dimensions of mussels found in the archaeological assemblages of southern Tierra del Fuego have previously been associated with variations influenced by microenvironmental factors. Notably, the variation in the representation of specimens smaller than 30 mm within shell deposits may be linked to the occasional use of different mussel harvesting techniques, influenced by topographical differences along the shoreline (Orquera and Piana 1999; 2001). In areas like Shamakush and Imiwaia, where the coast features a gentle, low-slope relief and predominantly soft substrates, the beaches differ—Shamakush has wide, sandy shores formed by river deposition, while Imiwaia's intertidal and subtidal bottom is muddier and more clayey. In both locations, mussels attach to scattered pebbles, allowing for manual extraction and enhancing selectivity in the size of the harvested specimens. As human selection likely targets larger mussels, experimental studies have demonstrated that this plucking method (Jones and Richman 1995) results in shell accumulations with an average size exceeding that of natural populations, with minimal representation of inadvertently transported individuals. Thus, these findings align with observations in the archaeological shell accumulations of these sites.

In contrast, in the Túnel-L. Packewaia locality, the coast is angular and features continuous rock extensions with sufficient support for numerous colonies of mussels (Orquera 2005). The rocky bottom deepens rapidly below the level of low tides, making size selection more challenging in this space and possibly requiring the use of tools to reach specimens submerged at certain depths. In these circumstances, mussels had to be lifted from the bottom in clusters that grouped specimens of all sizes, resulting in the incidental transport of very small specimens of minimal or no value for consumption. Hence, in the Túnel-L. Packewaia sites, on average, smaller valve dimensions and a higher proportion of suboptimal mussels were observed than in the previously mentioned locations. Additionally, from 1500 cal yr BP, this locality depicts an increase in the number of shell middens with high abundance indices, which might suggest a greater emphasis on the cluster collection method of mussels for reasons unknown up to this point. In Heshkaia, gently sloping coasts predominate, and there are many long pebble or sandy beaches interrupted by abrupt rocky stretches with mussel banks that are completely exposed during low tides. In line with this microenvironmental heterogeneity, the variable indices of valves with dimensions below 30 mm identified in the shell middens suggest the circumstantial use of both collection techniques.

Since shell middens represent time-averaged deposits, the scale at which ecological factors operate cannot fully explain the long-term size patterns identified in the area, nor does it exclude the influence of more general anthropogenic processes (Mannino and Thomas 2002; Faulkner 2013). In this regard, the question of whether the diachronic demographic and economic trends in the area triggered changes in foraging intensity of intertidal mussels is still up for debate and needs further exploration using indicators more directly linked to human impact. For example, the average ontogenetic age of exploited mollusks, unlike the size of the valves, is less affected by environmental variation (Claassen 1998; Mannino and Thomas 2002). Consequently, changes in the age profile of targeted specimens across different temporal periods would serve as a reliable means to determine whether the observed decline in average size during later stages genuinely signifies an increased emphasis on collection intensity.

Finally, it is also imperative to develop analytical and interpretative approaches that facilitate a more comprehensive understanding of the intricate interaction between human groups and malacological resources. Recently, multiple studies have argued that anthropogenic impact is a complex phenomenon whose effects are not always directional and do not mechanically translate into a decrease in the size of exploited specimens (Campbell 2008; Giovas et al. 2013; Thakar et al. 2017; Hausmann et al. 2021). In addition to considering the life history parameters of exploited species, this observation encourages the exploration of other aspects of mussel human exploitation also involved in the process of occupying the southern coast of Tierra del Fuego and potentially dynamic. In a context of economic intensification and resource diversification, it is plausible that human groups experimented with new modalities in mussel acquisition, reflected in changes in seasonal patterns of shellfish foraging or increased dependence on certain collection methods.

4.6 Conclusion

The metric studies conducted on the mussel shells recovered from archaeological shell middens in southern Tierra del Fuego have been crucial in understanding the role of shellfish resources in human subsistence during the Middle and Late Holocene. The reduced sizes of the archaeological specimens compared to natural populations not exposed to human predation underscore the economic significance of these animals and can be interpreted as the result of a long-term pattern of intensive localized shellfish exploitation by mobile hunter-gatherers who periodically altered size and possibly age composition of shell beds. Cultural and ecological factors restricted human impact to the local scale, with no effects on the metapopulation, maintaining ecological stability throughout the occupational sequence in the area. The rapid reproduction and fast growth of mussel species, especially in the case of *M. edulis*, along with their wide geographic distribution and proximity to larval refugia, helped counteract human depletion. Additionally, forager behavior played a role in against overexploitation through shifting to more productive locations (Orquera and Piana 2001; Mannino and Thomas 2002).

While no observable temporal changes in mussel abundance in shell middens denote long-term sustainability, diachronic tendencies in shell sizes suggest possible changes in the impact levels on the local population for at least 1500 cal yr BP. These changes certainly make sense in the overall context of economic reorganization and demographic increase in the area (Zangrando et al. 2020). However, the considerable variability in size frequency distributions among and within documented shell accumulations poses challenges to the interpretation of biometric information and the identification of diachronic patterns in shellfish foraging intensity. These challenges essentially arise from the intrinsic properties of the archaeological record, shaped by complex interactions between several natural and cultural factors influencing size representation in shell assemblages (Jerardino 1997; Claassen 1998; Campbell 2008; Hausmann et al. 2021). In relatively stable exploitation contexts where the resilience of consumed malacological species and human group spatial dynamics favor sustainable resource use over time, changes in collecting intensity might be more subtle and less evident than in cases of overexploitation typically considered in traditional archaeological literature. Therefore, this emphasizes the need to incorporate analytical approaches that allow for a clearer distinction between environmental and natural variables, such as, for example, research into age determinations and harvesting seasonal dynamics.

In conclusion, the variation in the biometric composition of archaeological shell middens in southern Tierra del Fuego raises new questions about other dimensions involved in human mussels' procurement contributing to variability in shell assemblages. As previously mentioned, the representation of small and immature mussels provides informative insights into collection techniques and their connection to the microenvironmental heterogeneity along the coastal area. In this respect, the review in this chapter provides an update on the potential offered by the study of malacological assemblages for the northern coast of the Beagle Channel and its relation to social and environmental factors operating at different scales.

References

Alunni DV, Tivoli AM, Zangrando AFJ (2020) Movilidad y explotación de recursos terrestres por cazadores-recolectores costeros de Bahía Moat, costa sureste de Tierra del Fuego, Argentina. Lat Am Antiq 31(3):1–19. https://doi.org/10.1017/laq.2020.27

Anderson AJ (1981) A model of prehistoric collecting on the rocky shore. J Archaeol Sci 8:109–120. https://doi.org/10.1016/0305-4403(81)90020-0

Bailey GN (1994) Weipa shell mounds: cultural or natural deposits? In: Sullivan M, Brockwell S, Webb A (eds) Archaeology in the North. Australian National University, North Australia Research Unity, Darwin, pp 107–129

Bala LO (1989) Biología y ecología del mejillón (*Mytilus edulis platensis*) en el golfo San José, provincia del Chubut. PhD Dissertation, Universidad Nacional de La Plata

Barcelo JA, Piana EL, Martinioni DR (2002) Archaeological spatial modeling: a case study from Beagle Channel (Argentina). British Archaeological Research International Series, Oxford

Bayne BL (1976) Marine mussels: their ecology and physiology. Cambridge University Press, New York

Benedetti-Cecchi L, Cinelli F (1997) Spatial distribution of algae and invertebrates in the rocky intertidal zone of the Strait of Magellan: are patterns general? Polar Biol 18(5):337–343. https://doi.org/10.1007/s003000050197

Blanchette CA, Broitman BR, Gaines SD (2006) Intertidal community structure and oceanographic patterns around Santa Cruz Island, CA, USA. Mar Biol 149(3):689–701. https://doi.org/10.1007/s00227-005-0239-3

Blanchette CA, Gaines SD (2007) Distribution, abundance, size and recruitment of the mussel, *Mytilus californianus*, across a major oceanographic and biogeographic boundary at Point Conception, California, USA. J Exp Mar Biol Ecol 340(2):268–279. https://doi.org/10.1016/j.jembe.2006.09.014

Botkin S (1980) Effects of human exploitation on shellfish populations at Malibu Creek, California. In: Earle TK, Christenson AL (eds) Modeling change in prehistoric subsistence economies. Academic Press, New York, pp 121–136

Braje TJ, Kennett DJ, Erlandson JM, Culleton BJ (2007) Human impacts on nearshore shellfish taxa: a 7000 year record from Santa Rosa Island California. Am Antiq 72(04):735–775. https://doi.org/10.2307/25470443

Campbell GE (2008) Beyond means to meaning: using distributions of shell shapes to reconstruct past collecting strategies. Environ Archaeol 13(2):111–121. https://doi.org/10.1179/174963108X343236

Campbell GE (2012) The archaeological potential of fragmentary remains of the Common Mussel, *Mytilus edulis*. Master Dissertation, University of Reading

Cannon A, Burchell M (2009) Clam growth-stage profiles as a measure of harvest intensity and resource management on the central coast of British Columbia. J Archaeol Sci 36(4):1050–1060. https://doi.org/10.1016/j.jas.2008.12.007

Castilla JC, Durán LR (1985) Human exclusion from the rocky intertidal zone of Central Chile: the effects on *Concholepas Concholepas* (Gastropoda). Oikos 45(3):391. https://doi.org/10.2307/3565575

Claassen CL (1998) Shells. Cambridge University Press, Cambridge

Codding BF, O'Connell JF, Bird DW (2014) Shellfishing and the colonization of Sahul: multivariate model evaluating the dynamic effects of prey utility, transport considerations and life-history on foraging patterns and midden composition. J Isl Coast Archaeol 9:238–252. https://doi.org/10.1080/15564894.2013.848958

Colonese AC, Camarós E, Verdún Castelló E, Giralt S, Rejas M (2011) Integrated archaeozoological research of shell middens: new insights into hunter-gatherer-fisher coastal exploitation in Tierra del Fuego. J Isl Coast Archaeol 6:235–254. https://doi.org/10.1080/15564894.2011.586088

Colonese AC, Verdún Castelló E, Álvarez M, Briz i Godino I, Zurro D, Salvatelli L (2012) Oxygen isotopic composition of limpet shells from the Beagle Channel: implications for seasonal studies in shell middens of Tierra del Fuego. J Archaeol Sci 39(6):1738–1748. https://doi.org/10.1016/j.jas.2012.01.012

Erlandson JM (1991) Shellfish and seeds as optimal resources: Early Holocene subsistence on the Santa Barbara coast. In: Erlandson JM, Colten RH (eds) Hunter gatherers of early holocene coastal California. University of California, Los Angeles, Cotsen Institute of Archaeology, pp 89–111

Erlandson JM, Braje TJ, Rick TC, Jew NP, Kennett DJ, Dwyer N, Ainis AF, Vellanoweth RL, Watts J (2011) 10,000 years of human predation and size changes in the owl limpet (*Lottia gigantea*) on San Miguel Island California. J Archaeol Sci 38(5):1127–1134. https://doi.org/10.1016/j.jas.2010.12.009

Erlandson JM, Rick TC, Braje TJ, Steinberg A, Vellanoweth RL (2008) Human impacts on ancient shellfish: a 10,000 year record from San Miguel Island, California. J Archaeol Sci 35:2144–2152. https://doi.org/10.1016/j.jas.2008.01.014

Faulkner P (2013) Investigating variability in the intensity of occupation and resource use. Life on the margins: an archaeological investigation of late holocene economic variability Northern Australia. ANU Press, Blue Mud Bay, pp 1–23

Flores C, Gayo EM, Salazar D, Broitman BR (2018) δ18O of *Fissurella maxima* as a proxy for reconstructing Early Holocene sea surface temperatures in the coastal Atacama desert (25°S). Palaeogeogr Palaeoclimatol Palaeoecol 499:22–34. https://doi.org/10.1016/j.palaeo.2018.03.031

Forcelli DO (2000) Moluscos magallánicos. Guía de moluscos de Patagonia y Sur de Chile. Vázquez Manzini Editores, Buenos Aires

Giovas CM, Clark M, Fitzpatrick SM, Stone J (2013) Intensifying collection and size increase of the tessellated nerite snail (*Nerita tessellata*) at the Coconut Walk site, Nevis, northern Lesser Antilles, AD 890–1440. J Archaeol Sci 40(11):4024–4038. https://doi.org/10.1016/j.jas.2013.05.008

Gordillo S (1991) Paleoecología de moluscos marinos del Holoceno Medio en Isla Gable, Canal Beagle (Tierra del Fuego, Argentina). Ameghiniana 28(1–2):127–133

Gordillo S, Brey T, Beyer K, Lomovasky BJ (2015) Climatic and environmental changes during the middle to late Holocene in southern South America: a sclerochronological approach using the bivalve *Retrotapes exalbidus* (Dillwyn) from the Beagle Channel. Quat Int 377:83–90. https://doi.org/10.1016/j.quaint.2014.12.036

Gosling EM (1992) Systematics and geographic distribution of Mytilus. In: Gosling EM (ed) The mussel Mytilus: ecology, physiology, genetics and culture. Developments in aquaculture and fisheries, vol 25. Elsevier, Amsterdam, pp 1–20

Gray AP, Seed R, Richardson CA (1997) Reproduction and growth of *Mytilus edulis chilensis* from the Falkland Islands. Sci Mar 61:39–48

Griffiths RJ (1981) Population dynamics and growth of the bivalve *Choromytilus meridionalis* (Kr.) at different tidal levels. Estuar Coast Shelf Sci 12(1):101–118. https://doi.org/10.1016/S0302-3524(81)80120-X

Groesbeck AS, Rowell K, Lepofsky D, Salomon AK (2014) Ancient clam gardens increased shellfish production: Adaptive strategies from the past can inform food security today. PLoS One 9(3). https://doi.org/10.1371/journal.pone.0091235

Hausmann N, Meredith-Williams M, Laurie E (2021) Shellfish resilience to prehistoric human consumption in the southern Red Sea: Variability in *Conomurex fasciatus* across time and space. Quat Int 584:20–32. https://doi.org/10.1016/j.quaint.2020.04.034

Hockey PAR, Bosman AL (1986) Nordic society Oikos man as an intertidal predator in Transkei: disturbance, community convergence and management of a natural food resource. Oikos 46:3–14. https://doi.org/10.2307/3565373

Iachetti C, Lovrich GA, Alder VA (2021) Temporal variability of the physical and chemical environment, chlorophyll and diatom biomass in the euphotic zone of the Beagle Channel (Argentina): evidence of nutrient limitation. Prog Oceanogr 195:102576. https://doi.org/10.1016/j.pocean.2021.102576

Jerardino A (1997) Changes in shellfish species composition and mean shell size from a Late Holocene record of the west coast of southern Africa. J Archaeol Sci 24(11):1031–1044. https://doi.org/10.1006/jasc.1997.0182

Jerardino A (2018) Shell fragmentation beyond screen-size and the reconstruction of intra-site settlement patterns: a case study from the west coast of South Africa. In: Giovas CM, LeFebvre MJ (eds) Zooarchaeology in practice: case studies in methodology and interpretation in archaeofaunal analysis. Springer, Cham, pp 151–171

Jerardino A, Navarro R (2008) Shell morphometry of seven limpet species from coastal shell middens in southern Africa. J Archaeol Sci 35:1023–1029. https://doi.org/10.1016/j.jas.2007.07.004

Jones TL, Richman JR (1995) On mussels: *Mytilus californianus* as a prehistoric resource. North Am Archaeol 16(1):33–58. https://doi.org/10.2190/G5TT-YFHP-JE6A-P2TX

Kilburn R, Rippey E (1982) Sea shells of Southern Africa. Macmillan, Johannesburg

Klein RG, Avery G, Cruz-Uribe K, Halkett D, Parkington JE, Steele T, Volman TP, Yates R (2004) The Ysterfontein 1 middle stone age site, South Africa, and early human exploitation of coastal resources. Proc Natl Acad Sci 101(16):5708–5715. https://doi.org/10.1073/pnas.0400528101

Langley S, Guzman L, Ríos C (1980) Aspectos dinámicos de *Mytilus chilensis* (Hupé, 1840) en el estrecho de Magallanes. I. Distribución, densidad y disposición espacial en el intermareal. An Del Inst La Patagon Ser Cs Nat 11:319–332

Lasiak T (1992) Contemporary shellfish-gathering practices of indigenous coastal people in Transkei: some implications for interpretation of the archaeological record. S Afr J Sci 88(1):19–28

Lasiak T (1991) The susceptibility and/or resilience of rocky littoral molluscs to stock depletion by the indigenous coastal people of Transkei, Southern Africa. Biol Conser 56(3):245–264. https://doi.org/10.1016/0006-3207(91)90060-M

Lasta ML, Ciocco NF, Bremec CS, Roux AM (1998) Moluscos bivalvos y gasterópodos. In: Boschi EE (ed) El Mar Argentino y sus recursos pesqueros, tomo 2. Los moluscos de interés pesquero. Cultivos y estrategias reproductivas de bivalvos y equinoideos. Instituto Nacional de Investigación y Desarrollo Pesquero, Secretaría de Agricultura, Ganadería, Pesca y Alimentación, pp 115–142
Mannino MA, Thomas KD (2002) Depletion of a resource? The impact of prehistoric human foraging on intertidal mollusc communities and its significance for human settlement, mobility and dispersal. World Archaeol 33(3):452–474. https://doi.org/10.1080/00438240120107477
Mannino MA, Thomas KD (2001) Intensive mesolithic exploitation of coastal resources? Evidence from a shell deposit on the Isle of Portland (Southern England) for the impact of human foraging on populations of intertidal rocky shore molluscs. J Archaeol Sci 28:1101–1114. https://doi.org/10.1006/jasc.2001.0658
Martinoli MP (2018) Modalidades de explotación, procesamiento y consumo de pinnípedos en la margen meridional de Tierra del Fuego. PhD Dissertation, Universidad de Buenos Aires
McDonald JH, Seed R, Koehn RK (1991) Allozymes and morphometric characters of three species of *Mytilus* in the Northern and Southern Hemispheres. Mar Biol 111(3):323–333. https://doi.org/10.1007/BF01319403
Menge BA, Chan F, Lubchenco J (2008). Response of a rocky intertidal ecosystem engineer and community dominant to climate change. Ecol Lett 11(2):151–162. https://doi.org/10.1111/j.1461-0248.2007.01135.x
Mestre NC, Thatje S, Tyler PA (2009) The ocean is not deep enough: Pressure tolerances during early ontogeny of the blue mussel *Mytilus edulis*. Proc R Soc B Biol Sci 276(1657):717–726. https://doi.org/10.1098/rspb.2008.1376
Morán GA, Bayer S, Beierlein L, Martínez JJ, Benitez-Vieyra S, Mackensen A, Brey T, Gordillo S (2021) Late Holocene *Ameghinomya antiqua* shells from the Beagle Channel: a multi-proxy approach to palaeoenvironmental and palaeoclimatic reconstruction. Palaeogeogr Palaeoclimatol Palaeoecol 578:110574. https://doi.org/10.1016/j.palaeo.2021.110574
Moreno CA, Sutherland JP, Jara HF (1984) Man as a predator in the intertidal zone of Southern Chile. Oikos 42(2):155. https://doi.org/10.2307/3544787
Nicastro A, Surge D, Briz i Godino I, Álvarez M, Schöne BR, Bas M (2020) High-resolution records of growth temperature and life history of two *Nacella* limpet species, Tierra del Fuego, Argentina. Palaeogeogr Palaeoclimatol Palaeoecol 540(2):1–59. https://doi.org/10.1016/j.palaeo.2019.109526
Obelic B, Álvarez A, Argullós J, Piana EL (1998) Determination of water paleotemperature in Beagle Channel (Argentina) during the last 6000 yr through stable isotope composition of Mytilus edulis shells. Quat South Am Antart Penins 11:47–71
Oehrens Kissner EM, Kroeck MA (2005) Estimación de la talla de primera madurez sexual del mejillón, *Mytilus edulis platensis* (Mollusca: Bivalvia) en el Golfo San Matías. IBMP Ser Public IV:21–44
Ojeda J, Rosenfeld S, Marambio J, Rozzi R, Mansilla A (2014) Patrones estacionales y espaciales de la diversidad de moluscos intermareales de bahía Róbalo, canal Beagle, Reserva de la Biosfera Cabo de Hornos Chile. Rev Biol Mar Oceanogr 49(3):493–509. https://doi.org/10.4067/S0718-19572014000300007
Orquera LA (1999) El consumo de moluscos por los canoeros del extremo sur. Relac Soc Arg Antropol XXIV:307–327
Orquera LA (2005) Mid-Holocene littoral adaptation at the southern end of South America. Quat Int 132:107–115. https://doi.org/10.1016/j.quaint.2004.07.019
Orquera LA, Piana EL (1989–90) La formación de los montículos arqueológicos de la región del canal Beagle. Runa XIX:59–82. https://doi.org/10.34096/runa.v19i1.4271
Orquera LA, Piana EL (1992) Un paso hacia la resolución del palimpsesto. In: Borrero LA, Lanata JL (eds) Análisis espacial en la arqueología patagónica. Ayllu SRL, Buenos Aires, p 21–52
Orquera LA, Piana EL (1991) La formación de los montículos arqueológicos de la región del canal Beagle. Runa 19:59–82. https://doi.org/10.34096/runa.v19i1.4271

Orquera LA, Piana EL (1993–94) Lancha Packewaia: actualización y rectificaciones. Relac Soc Arg Antropol XIX:325–362
Orquera LA, Piana EL (1999) Arqueología de la región del canal Beagle (Tierra del Fuego, República Argentina). Sociedad Argentina de Antropología, Buenos Aires
Orquera LA, Piana EL (2000) Composición de conchales de la costa del canal Beagle (Tierra del Fuego, República Argentina)—primera parte. Relac Soc Arg Antropol XXV:249–274
Orquera LA, Piana EL (2001) Composición de conchales de la costa del canal Beagle (Tierra del Fuego, República Argentina)—segunda parte. Relac Soc Arg Antropol XXVI:345–368
Orquera LA, Piana EL (2009) Sea nomads of the Beagle Channel in Southernmost South America: over six thousand years of coastal adaptation and stability. J Isl Coast Archaeol 4:61–81. https://doi.org/10.1080/15564890902789882
Orquera LA, Piana EL (2020) Calibración radiocarbónica en la región del canal Beagle. Relac Soc Arg Antropol 45(1):201–221
Page HM, Hubbard DM (1987) Temporal and spatial patterns of growth in mussels *Mytilus edulis* on an offshore platform: relationships to water temperature and food availability. J Exp Mar Bio Ecol 111(2):159–179. https://doi.org/10.1016/0022-0981(87)90053-0
Parkington J, Brand R (2020) Shell middens and coastal prehistory. Oxf Res Encycl Anthropol. https://doi.org/10.1093/acrefore/9780190854584.013.48
Pérez SA (2020) La explotación de moluscos intermareales durante el Holoceno tardío en la localidad de Heshkaia, bahía Moat, sur de Tierra del Fuego. Intersecc Antropol 21(1):43–56. https://doi.org/10.37176/iea.21.1.2020.529
Piana EL, Orquera LA (2010) Shell midden formation at the Beagle Channel (Tierra del Fuego, Argentine). In: Calado D, Baldia M, Boulanger M (eds) Monumental questions: prehistoric megaliths, mounds and enclosures british archaeological reports international series 2122. Archaeopress, Oxford, pp 263–273
Raab LM (1992) An optimal foraging analysis on San Clemente Island California. J Ethnobiol 12(1):63–80
Rick TC, Robbins JA, Ferguson KM (2006) Stable isotopes from marine shells, ancient environments, and human subsistence on middle Holocene Santa Rosa island, California, USA. J Isl Coast Archaeol 1(2):233–254. https://doi.org/10.1080/15564890600935480
Rosenfeld S, Marambio J, Aldea C (2013) Comparación de ensambles de moluscos en dos sustratos intermareales de la cuenca central del Estrecho de Magallanes. Amici Moluscarum 21(2):7–18
Seed R (1969a) The ecology of *Mytilus edulis L.* (Lamellibranchiata) on exposed rocky shores I. Breeding and Settlement. Oecologia 3:277–316. https://doi.org/10.1007/BF00390380
Seed R (1969b) The ecology of *Mytilus edulis L.* (Lamellibranchiata) on exposed rocky shores II Growth and Mortality. Oecologia 3:317–350. https://doi.org/10.1007/BF00390381
Seed R (1968) Factors influencing shell shape in the mussel *Mytilus edulis*. J Mar Biol Assoc UK 48:561–584. https://doi.org/10.1017/S0025315400019159
Seed R, Suchanek TH (1992) Population and community ecology of Mytilus. In: Gossling ED (ed) The mussel Mytilus: ecology, physiology, genetics and culture. Elsevier, New York, pp 87–170
Silva MR, Calvo J (1995) Influencia ambiental sobre la morfología valvar de *Mytilus edulis chilensis* del Canal Beagle. In: Actas del IV Congreso Latinoamericano de Ciencias del Mar, Mar del Plata, pp 186
Suchanek TH (1981) The role of disturbance in the evolution of life history strategies in the intertidal mussels *Mytilus edulis* and *Mytilus californianus*. Oecologia 50:143–152. https://doi.org/10.1007/BF00348028
Thakar HB, Glassow MA, Blanchette C (2017) Reconsidering evidence of human impacts: implications of within-site variation of growth rates in *Mytilus californianus* along tidal gradients. Quat Int 427:151–159. https://doi.org/10.1016/j.quaint.2015.10.018
Tortorelli M del C (1987) Contribución al estudio de los ciclos reproductivos del mejillón patagónico *Mytilus chilensis Hupé*, y de la cholga *Aulacomya ater* (Molina) en el Canal Beagle. PhD Dissertation, Universidad de Buenos Aires

Trancart M (1978) Biologie et possibilites d'explotation de *Mytillus platensis* (d'Orb) dans le Golfe San José, Péninsule Valdés, Argentine. PhD Dissertation, Université D' Aix-Marseille II

Verdún Castelló E (2011) El consum de mol·luscs en societats caçadores-recol·lectores de Tierra del Fuego (Argentina). PhD Dissertation, Universidad Autónoma de Barcelona

Verdún Castelló E (2014) Application of biometric analyses on shell middens of hunter-fisher-gatherer societies of Tierra del Fuego (Argentina). Archaeofauna 23:223–238

Vinuesa JH (1979) Ciclo gonadal y primera madurez sexual del mejillón patagónico *Mytilus edulis chilensis Hupe* de Pto. Deseado. Physis 38(95):35–47

Wanamaker AD, Kreutz KJ, Borns HW, Introne DS, Feindel S, Funder S, Rawson PD, Barber BJ (2007) Experimental determination of salinity, temperature, growth, and metabolic effects on shell isotope chemistry of *Mytilus edulis* collected from Maine and Greenland. Paleoceanography 22(2):1–12. https://doi.org/10.1029/2006PA001352

Waselkov GA (1987) Shellfish gathering and shell midden archaeology. Adv Archaeol Method Theory 10:93–210. https://doi.org/10.1016/B978-0-12-003110-8.50006-2

Whitaker AR (2008) Incipient aquaculture in prehistoric California?: Long-term productivity and sustainability versus immediate returns for the harvest of marine invertebrates. J Archaeol Sci 35:1114–1123. https://doi.org/10.1016/j.jas.2007.08.005

Whitaker AR, Byrd BF (2014) Social circumscription, territoriality, and the Late Holocene intensification of small-bodied shellfish along the California Coast. J Isl Coast Archaeol 9(2):150–168. https://doi.org/10.1080/15564894.2014.881934

Zaixso H, Ré M, Morsán E (2015) Moluscos costeros de interés económico actual o potencial. In: Zaixso H, Boraso A (eds) La Zona Costera Patagónica Argentina, vol I: Recursos Biológicos Bentónicos. Universidad Nacional de la Patagonia SJB—EDUPA Editors, pp 128–308

Zangrando AF (2009) Historia evolutiva y subsistencia de cazadores recolectores marítimos de Tierra del Fuego. Sociedad Argentina de Antropología, Buenos Aires

Zangrando AF (2010) Coastal archaeology and hunter-gatherers in southeastern Tierra del Fuego. J Isl Coast Archaeol 5(2):588–291. https://doi.org/10.1080/15564894.2010.487360

Zangrando AF, Panarello H, Piana EL (2014) Zooarchaeological and stable isotopic assessments on pinniped-human relations in the Beagle Channel (Tierra del Fuego, southern South America). Int J Osteoarchaeol 24(2):231–244. https://doi.org/10.1002/oa.2352

Zangrando AF, Pinto Vargas G, Tivoli AM (2017) Decreased foraging return in shellfishing? Species composition and shell size of blue mussel (Mytilus edulis) from a Late Holocene site of the South Coast of Tierra del Fuego. Quat Int 427:160–169. https://doi.org/10.1016/j.quaint.2015.12.077

Zangrando AF, Vázquez M, Tessone A (2011) Los cazadores-recolectores del extremo oriental fueguino. Arqueología de Península Mitre e Isla de los Estados. Sociedad Argentina de Antropología, Buenos Aires

Zangrando AFJ, Tivoli A, Vázquez M, Alunni D, Kochi S, Martinoli MP, Pérez S, Tessone A, Butto A, Pinto Vargas G (2022) Arqueología de cazadores-recolectores, paleoecología y poblamiento de la costa sur de Tierra del Fuego. Poster presented at the Congreso Internacional Magallanes 2020–2022. 500 años de la primera circunnavegación de la Tierra, Punta Arenas

Zangrando AFJ, Tivoli AM, Alunni DV, Pérez SA, Martinoli MP, Vargas GP (2020) Exploring shell midden formation through tapho-chronometric tools: a case study from Beagle Channel, Argentina. Quat Int 584:33–43. https://doi.org/10.1016/j.quaint.2020.04.050

Zuschin M, Stachowitsch M, Stanton RJ (2003) Patterns and processes of shell fragmentation in modern and ancient marine environments. Earth Sci Rev 63:33–82. https://doi.org/10.1016/S0012-8252(03)00014-X

Chapter 5
The Ecological Niche of Hunter-Gatherers from the Southern Coast of Tierra del Fuego: An Isotopic Perspective to Connect Archaeology and Historical Ecology

Sayuri Kochi, Augusto Tessone, and Atilio Francisco J. Zangrando

Abstract A discussion about subsistence strategies and ethnic identities inspired the first study of diets with carbon and nitrogen stable isotopes in Tierra del Fuego. From this onwards, stable isotope analyses on human remains deepened the knowledge about the spatial variation of native diets. Furthermore, they proposed regional continuities where written sources projected discrete cultural boundaries between prehistoric groups of Beagle Channel and Península Mitre. The isotopic perspective also contributed to the reconstruction of ancient food webs previous to the industrial exploitation of pinnipeds, fisheries, and the introduction of livestock in southern South America. In this chapter, we synthesize 30 years of stable isotope analyses on these topics, focusing on the construction of local isotopic ecology, and the scopes and limitations of human paleodietary reconstructions in the southern coast of Tierra del Fuego. From these works, we highlight the role of human populations as consumers connecting the terrestrial and marine food webs, which gave us a better understanding of the flexibility of native groups facing major challenges such as industrial exploitation of fauna and contact processes during the Late Holocene.

S. Kochi (✉)
Department of Pre- and Protohistory, Institute of Ancient Studies, Johannes Gutenberg University Mainz, GermanyMainz
e-mail: ksayuri@uni-mainz.de

S. Kochi · A. Tessone
Instituto de Geocronología y Geología Isotópica (CONICET/UBA), CABA, Argentina
e-mail: atessone@ingeis.uba.ar

A. F. J. Zangrando
CADIC-CONICET, Ushuaia, TDFeIAS, Argentina

FFyL-UBA, CABA, Argentina

A. F. J. Zangrando
e-mail: fzangrando@conicet.gov.ar

A. F. J. Zangrando et al. (eds.), *Archaeology of the South Coast of Tierra del Fuego*, The Latin American Studies Book Series, https://doi.org/10.1007/978-3-032-09487-2_5

5.1 Introduction

Stable isotope analyses share a deep bond with coastal archaeology. As a methodology integrated with other evidence of the material record, they contribute to rethinking the spatial scales of subsistence and mobility in coastal hunter-gatherers (Borrero and Barberena 2006). For example, early isotopic works challenged previous models on subsistence strategies (Chisholm et al. 1982; Sealy and van der Merwe 1985). Moreover, they can provide environmental information from ubiquitous components of coastal archaeological sites, such as mollusk shells (Leng and Lewis 2016). Fuegian archaeology is not an exception in this trend: Panarello (1987) published a pioneer work in Argentina, reconstructing seawater paleotemperatures from $\delta^{18}0$ in archaeological shellfish. Later, Yesner et al. (1991) presented the first values of $\delta^{13}C_{apatite}$ and $\delta^{15}N$ measured on human bone remains in Argentina—alongside $\delta^{13}C_{collagen}$—to discuss prehistoric diets, subsistence strategies, and ethnic identities in Tierra del Fuego. From this onwards, stable isotope analyses on human remains deepened the knowledge about the spatial variation of native diets (Orquera and Piana 1996; Borrero et al. 2001; Guichón et al. 2001; Yesner et al. 2003; Barberena 2004). To characterize the flexibility of native lifestyles, it is necessary to switch between different scales and analytical units of the archaeological record (Barberena et al. 2004; Tessone et al. 2011), and this is one of the major contributions of isotopic analyses in the southern coast of Tierra del Fuego: the information recorded in the lifespan of the individuals suggests regional continuities where written sources projected discrete cultural boundaries between prehistoric groups (Tessone et al. 2003; Yesner et al. 2003; Panarello et al. 2006; Tessone et al. 2011).

Thinking about these human groups as consumers, linking marine and terrestrial food webs (Maschner et al. 2009), inspired a second set of isotopic works: the reconstruction of ancient food webs previous to the industrial exploitation of pinnipeds (Zangrando et al. 2014a, b; Nye et al. 2020), fisheries (Zangrando et al. 2016), and the introduction of livestock (Kochi et al. 2020) in the southern tip of South America. The goal was to generate a frame of reference for classical models from the optimal foraging theory, to evaluate changes and continuities in marine hunter gatherer lifestyles. Moreover, we aimed to disentangle the diversity subsumed in the "marine diet" category (Kochi et al. 2018) with analytical advances in paleodietary reconstructions, such as Bayesian mixing models (Kochi 2017). We gained a better understanding of the ecological niches of species in the long term and their responses to major human impacts. In turn, we also identified the limitations of paleodietary interpretations with bulk stable isotope analyses, related with the characteristics of the local isotopic ecology. In this sense, we observe another link between stable isotope analyses and coastal archaeology: new questions about past societies, and the desire to extend the limits of paleodietary reconstructions are demanding advances from the biogeochemical field, such as compound-specific stable isotope analyses (Webb et al. 2018).

It is possible to group a third set of isotopic studies in the southern coast of Tierra del Fuego, where the assessment of the macronutrient content of native diets enriched further discussions about topics such as health (Schinder and Guichón 2003), spatial diversity in diets (Barberena 2004), postcontact continuities in traditional lifestyles (Tafuri et al. 2017), and scheduled handling of fats in a high-latitude environment (Alunni et al. 2021). Various analytical models were used in paleodietary reconstructions (Ambrose et al. 1997; Froehle et al. 2010), revealing their strengths and limitations. On the other hand, these topics serve as a reminder that stable isotopes most accurately characterize the final phase of foodways—consumption (Goody 1982) and more strictly speaking, the assimilation. The richest insight into the past is achieved when considering the previous phases of procurement, distribution, preparation, and disposal of food from other lines of archaeological evidence. For example, long-term isotopic information recorded on human bones, discussed with analyses on extractive techniques of bone grease, revealed a flexible processing of guanacos while securing the provision of fat throughout the annual cycle (Alunni et al. 2021).

In this chapter we synthesize 30 years of stable isotope analyses in the archaeology of the southern coast of Tierra del Fuego, focusing on these main themes: (1) spatial variation of native diets, historical ethnic boundaries, and archaeological continuities; (2) an isotopic perspective connecting archaeology and historical ecology; (3) linking nutritional aspects of diet with lifestyles in the past.

5.2 First Discussions with Stable Isotopes in Tierra del Fuego

The first application of stable carbon isotopes in the Beagle Channel was related to the determination of the local marine reservoir effect (ΔR) in the Middle Holocene, from multiple samples of Túnel I site shell midden (Albero et al. 1986). This work was concerned with spatial and temporal variation in ΔR, for calibration of ^{14}C age in archaeological remains of marine origin. In addition, it is also crucial to calibrate ^{14}C age in humans who consumed a mix of marine and terrestrial resources (Arneborg et al. 1999).

However, the interest in research with stable isotopes shifted to characterization of human diets for subsistence models. Along the southern coast of Isla Grande de Tierra del Fuego, native populations thrived as marine hunter-gatherers (Orquera and Piana 1999a, 2009). Since the end of nineteenth century, written sources called these inhabitants of the Beagle Channel Yagan/Yámana[1] (Bridges 1886 in Orquera and Piana 1999b). They were depicted traveling in canoes and capturing diverse fauna, including pinnipeds, fish, birds, guanacos, and mollusks. The French Scientific Mission to Cape Horn (1882–1883) spread to a wider public a map with a tripartite ethnic division of the Southern tip of South America (Fig. 5.1). Alongside Yagans,

[1] The first term of the pair is the current denomination of the local two Indigenous Communities, Yagan Paiakoala in Ushuaia and Yagan Bahía Mejillones in Puerto Williams. The second term, also registered by etnohistorical accounts, was used in our previous publications. For a detailed history of the ethnonym, we refer to Orquera and Piana (1999b): 77–79.

the Kawésqar/Alakaluf[2] of the Southwestern archipelago were classified as maritime groups; since the early nineteenth century, they were differentiated by linguistic observations of Fitz-Roy (1839). The Selk'nam/Ona inhabiting the northern steppe of the main island were characterized as terrestrial hunter-gatherers.

The Anglican missionary Thomas Bridges noted linguistic variations in "eastern onas" (Bridges 1897 in Saletta 2017). From the twentieth century onwards, native groups in Península Mitre—the eastern tip of the island—were identified as Haush (Coiazzi 1914) and ethnographies presented a map with four groups in Tierra del Fuego (Fig. 5.1, right; Lothrop 1928). But who were these groups? For example, Furlong (1917: 181) wrote: "The Haush were not a strong foot people, nor a canoe people, though there seems to be evidence that during the quieter season of the year some of them used canoes".

For such a "small mysterious tribe (…) formerly inhabitants of the eastern tip of Tierra del Fuego" (Lothrop 1928: 106), Hernán Vidal analyzed written records of contacts dating back to sixteenth century (Vidal 1985, 2011). From an ethnohistorical perspective, he highlighted diachronic changes in their characterization. However, the works of ethnographers Martin Gusinde (1986 [1937]) and Anne Chapman (Chapman and Hester 1973), established the most influential viewpoint: that the populations who inhabited Península Mitre were biologically and culturally related to the Selk'nam.

This was the background of the article published by Yesner et al. (1991, 2003), "Stable isotope analysis of human bone and ethnohistoric subsistence patterns in Tierra del Fuego." It included 14 individuals, estimated to be from the last 1500 years and the areas associated with the ethnographical maps: the north of the Tierra del Fuego Island (the Selk'nam), the northern littoral of Península Mitre (the Haush), and the Beagle Channel, Hoste and Navarino Islands (the Yagan). In the $\delta^{13}C_{col}$ versus $\delta^{15}N$ biplot, the northern individuals form a cluster identified with a predominantly terrestrial diet. On the other hand, most individuals from the Península Mitre and the Beagle Channel are clustered toward the most positive δ-values, reflecting a higher consumption of marine resources. The results confirmed the characterization of these groups registered by the written sources, but questioned two aspects: (1) the Haush group, generally affiliated with terrestrial hunter-gatherers, were isotopically similar to Yagan individuals; (2) one individual of the latter group had an unexpected contribution of terrestrial resources, and suggested that the importance of marine prey in the ethnohistorical accounts, such as shellfish, could be related to the impact of the European colonization. The conclusion mentioned the potential to discuss long-term averaging isotopic data, with previous analyses focusing on site-specific archaeofaunal data. Later, Orquera and Piana (1996) published a $\delta^{13}C_{col}$ value measured on a neonate deposited in a shell midden, as part of a site report. From onwards, the research projects in Tierra del Fuego integrated isotopic analyses as an indispensable methodology.

---

[2] There are seven Kawésqar communities registered in Chile, Región de Magallanes (FUCOA 2014), while the term Alacalufe was derived from the narrations of Fitz-Roy (1839: 135).

Fig. 5.1 Ethnographic maps showing the distribution of native groups in Tierra del Fuego. Left: Hyades and Deniker (1891) mentioned the Yagan, Alakaluf and Ona. Right: Lothrop (1928) added a fourth group in the eastern tip of the island, the Haush

Subsequent works focused mostly on the spatial distribution of isotopic data, the dietary composition and its relation with the subsistence systems of hunter-gatherers (Fig. 5.2; Guichón et al. 2001; Schinder et al. 2003; Barberena 2004). In this sense, chronological trends were assessed for a bigger spatial scale, including the Fuegian Archipelago and Southern continental Patagonia (Borrero et al. 2001). There are no human remains from the southern coast of Tierra del Fuego dated to the Middle Holocene. The closest individual consuming a marine diet, dated to 6290 ± 50 ^{14}C BP (7000–7300 cal yr BP), was found in the Punta Santa Ana 1 site on the Magellan Strait (Borrero et al. 2001; San Román 2010). However, the archaeological projects in the Beagle Channel have already registered at least seven thousand years of coastal adaptations for these populations (Orquera and Piana 1999b, 2009), and isotopic works expanding the size of human samples confirmed the predominance of marine resources in long-term diets along the southern coast of Tierra del Fuego (Fig. 5.3; Tessone et al. 2003; Panarello et al. 2006). In turn, these dietary patterns were coherent with other lines of evidence connecting both Beagle Channel and Península Mitre, such as the chronological trends in bone technology and the distribution of radiocarbon dates (Zangrando et al. 2009; Vázquez et al. 2011).

In summary, this first decade of isotopic analyses in Tierra del Fuego was grounded on long-term research projects that have been developed theoretical models to explain the deep history of the native populations and the variability of the archaeological record. Directly or indirectly, "[h]ow old are the cultural configurations which came to be known as Aónikenk, Selk'nam, Kawéskar and Yámana?" (Borrero 1997: 60) was among the central topics in this set of isotopic works. In this sense, the local isotopic information renewed these classical debates of hunter-gatherer archaeology, such as the spatial scale of land-use cycles (Binford 2002), the flexibility of subsistence systems and the relationship between archaeology and ethnography (Wobst 1978).

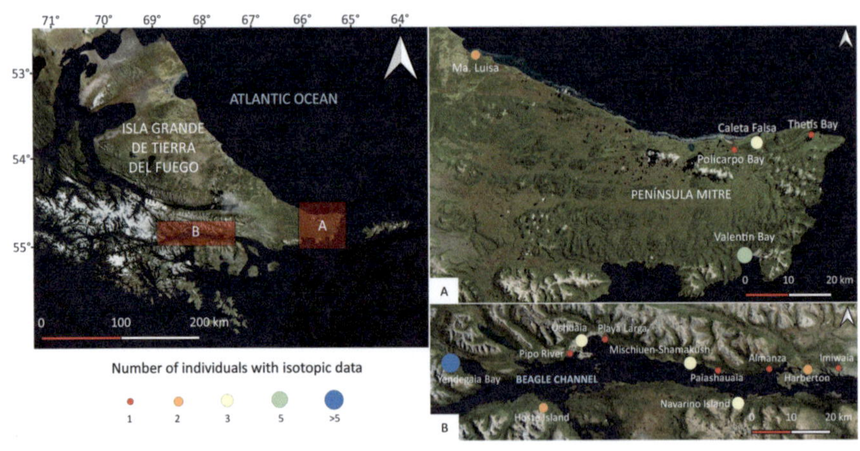

Fig. 5.2 Individuals from the Beagle Channel and Península Mitre with $\delta^{13}C_{col}$, $\delta^{15}N$, and/or $\delta^{13}C_{apa}$ values

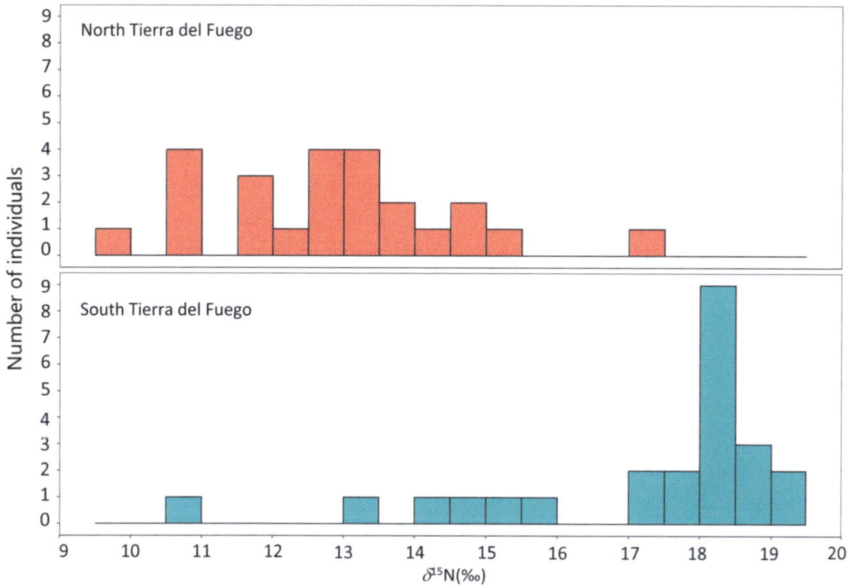

Fig. 5.3 Distribution of $\delta^{15}N$ values from Isla Grande de Tierra del Fuego, between 1700 cal. AP-1850 AD. References: Northern Tierra del Fuego (Borrero et al. 2001; Guichón et al. 2001; Yesner et al. 1991; Zangrando et al. 2003; Santiago et al. 2011). Southern Tierra del Fuego (Yesner et al. 1991; Guichón et al. 2001; Orquera and Piana 1996; Tessone et al. 2003; Panarello et al. 2006; Suby et al. 2011; Alfonso-Durruty et al. 2015; Kochi 2017)

5.3 Archaeology, Food Webs, and Historical Ecology

Shell middens are conspicuous features of the coastal landscapes in the Beagle Channel with exceptional bone preservation. The earliest sites associated with littoral adaptations were dated toward 7300 cal yr BP, and ring-shaped shell middens resulted from repeated occupations at the same place (Orquera and Piana 1999a, 2009; Zangrando 2018). From these sequences, zooarchaeological studies proposed an expansion of foraging habitats; as well as an increase of labor investment into fishing activities and the capture of birds in offshore waters during the Late Holocene (Zangrando 2009; Tivoli and Zangrando 2011). The exploitation of pinnipeds—a high-ranked resource in the area—also changed over time, as their abundance in zooarchaeological assemblages and the size of adults captured decreased throughout the Middle and Late Holocene (Tivoli and Zangrando 2011; Zangrando et al. 2014a; Martinoli 2018).

The use of different ecological zones for foraging activities was based mainly on the modern distribution of fauna in southern South America, integrating the expectations about the accessibility for human foragers. However, as the current faunal ecology was also a historical result of past human activities—both of small-scale economies and industrial overexploitation—the discussion about human-animal relations in the past needed at least a comparison between the modern and past distribution of fauna, as well as an appropriate environmental baseline extending back in time (Zangrando et al. 2014b).

At first, $\delta^{13}C$ analyses on local fauna have been carried out to estimate the ΔR as mentioned (Albero et al. 1986), but mainly to interpret the human diets (Orquera and Piana 1996). The same species but from other areas were also used as marine-terrestrial end members (Borrero et al. 2001; Barberena 2004; Panarello et al. 2006). In contrast, it is possible to recognize a second set of isotopic works on the southern coast of Tierra del Fuego with a zooarchaeological perspective, focusing on the paleoecological niche of key taxa for native populations and its changes associated with human subsistence patterns (Zangrando et al. 2022). The main principle is that carbon and nitrogen isotopes are intrinsic markers in the animal tissue (Hobson 2007), recording information on their habitat and feeding behavior (Newsome et al. 2007; Layman et al. 2012). The resolution of this information is dependent on the type of tissue analyzed, and the spatial distribution of isotopes in the environment. Nearshore primary producers are more enriched in ^{13}C than offshore ones (Casey and Post 2011; Clementz and Koch 2001; France 1995). For the nitrogen stable isotope values, there were expected variations related to the trophic position of the species (Minagawa and Wada 1984; Hedges and Reynard 2007). Moreover, $\delta^{15}N$ baseline values in this region (for example, particulate organic matter, phyto- and zooplankton, and sediment) change from inshore to offshore marine areas, which could be reflected in the isotopic values of past upper consumers (Riccialdelli et al. 2010, 2017).

Long-term human predation of pinnipeds before industrial exploitation could have originated a resource depression in the Beagle Channel (Zangrando 2009; Tivoli and Zangrando 2011); this hypothesis, formulated from the composition of zooarchaeological assemblages, was tested from stable isotope analyses as an independent line of inquiry. From $\delta^{13}C_{col}$ values on bone, it was suggested behavioral changes in female adults of South American fur seals (*Arctocephalus australis*) through the Middle and Late Holocene, because they avoided nearshore areas of the Beagle Channel for breeding (Zangrando et al. 2014a, b). The following work of Nye et al. (2020) included both fur seals and sea lions (*Otaria flavescens*) from the southern coast of Tierra del Fuego and also observed $\delta^{13}C_{col}$ and $\delta^{15}N$ variations over time.

The assessment of fishing activities in the past also relies on the modern distribution of species—for example, relative to the distance of the coast—or ontogenetic behavioral patterns. Despite the impact of modern commercial fisheries since the mid-twentieth century, the $\delta^{13}C$ values support the offshore character of southern hake (*Merluccius* sp.) and snoek (*Thyrsites atun*) in comparison with Patagonian grenadier (*Macruronus magellanicus*) and coastal fish taxa in prehistoric times (Zangrando et al. 2016). In this sense, isotopic analyses also supported the hypothesis about the expansion of fishing ranges toward the last 1500 cal yr BP.

In addition to univariate analyses, new analytical tools such as the isotopic niches (Newsome et al. 2007; Layman et al. 2012) helped to model expectations and standardize comparisons about the ecological niches of species and populations. Shortly, the position of individuals according to their isotopic values in a *δ*-biplot forms the isotopic niche at the population level. While the isotopic niches do not directly represent the ecological niche, it allows us to make inferences about the latter because the axes of isotopic markers ($\delta^{13}C$, $\delta^{15}N$, $\delta^{18}O$, etc.) supply information about resource/habitat use (Newsome et al. 2007). The derived metrics such as corrected standard ellipses areas (SEAc in ‰^{2}) can be interpreted as a proxy for interindividual variability in diet and/or habitat, and the Bayesian approach (SEAb) facilitates comparisons of niches (Jackson et al. 2011). For example, the biggest isotopic niche size in historic samples of fur seals was attributed to less interspecific and intraspecific competition after a significant population bottleneck in the nineteenth and twentieth centuries (Nye et al. 2020).

In terrestrial environments, the focus is on the guanaco (*Lama guanicoe*): a medium-sized herbivore hunted by foragers throughout the Holocene in southern South America. In the Isla Grande de Tierra del Fuego, the highest environmental contrast is generated by the orographic effect: dry steppe habitats to the north of the Fuegian Andes and forests to the south (Tuhkanen 1992). The isotopic niches of archaeological guanacos from these areas were clearly separated in the δ-biplot. Moreover, they showed that populations of southern guanacos occupied and fed in forested areas since the Middle Holocene. In conclusion, guanacos had an extensive use of the forest previous to the competitive displacement by introduced livestock in the nineteenth–twentieth century (Kochi et al. 2020).

Thinking about humans as another consumer of the food web, it was reasonable to look beyond the "marine diet" category and consider different combinations of marine resources in the long-term diet. Shellfishes, fishes, birds, and pinnipeds can have different isotopic values in $\delta^{13}C$ and $\delta^{15}N$ depending on their habitat and/or trophic positions (Richards and Hedges 1999). As they require different amounts of labor in their procurement and processing (Orquera and Piana 1999a; Zangrando 2009), a more detailed paleodietary reconstruction at the interindividual level would enrich the characterization of hunter-gatherers exploiting both nearshore and offshore environments. These interpretations are dependent on the characteristics of the local isotopic ecology, defined as "the isotopic values for the vegetal and animal foods potentially available for forager populations in a given environment" (Barberena and Tessone 2018). For example, if the intragroup isotopic variation in pinnipeds is larger than intergroup variation—compared with birds, or fishes—or if there is little variation between source groups, paleodietary estimations in human consumers with mathematical mixing models would have less taxonomical resolution than zooarchaeological analysis (Phillips et al. 2014).

Therefore, to comprehend better the scopes and limitations of human paleodietary reconstructions with stable isotopes, we measured the $\delta^{13}C$ and $\delta^{15}N$ bone collagen values on thirteen taxa from zooarchaeological assemblages of the Beagle Channel (Fig. 5.4; Kochi et al. 2018). This revealed a tight clustering of marine resources in the isotopic δ-space. When corrected by bone collagen-muscle offset, the marine vertebrate cluster has a wide gap compared with modern mussels (the latter corrected for the oceanic Suess effect; Misarti et al. 2009). Terrestrial resources, including both guanacos and plants, are the most negative end-members in the positive linear distribution of resources. In conclusion, bulk $\delta^{13}C$ and $\delta^{15}N$ analyses on human bone collagen would not be the best method to detect the regular consumption of pelagic animals as intended. Instead, the knowledge on the isotopic ecology of Beagle Channel cautions about multiple combinations of resources in human diets resulting in similar isotopic values. In this context, seabirds and fur seals may occupy similar trophic positions in the food web and it is difficult to differentiate diets relying on a high-ranked prey or a combination of multiple smaller animals.

Bayesian mixing models are an analytical tool for paleodietary reconstructions which need adequate knowledge of isotopic ecology (Phillips et al. 2014). They estimate what would have been the most likely combination of resources assimilated and their proportion, given the measured isotopic composition of the consumer. The results are expressed as probability distributions. These models also work with underdetermined systems: when there are n isotopic markers available for analyses (i.e., $n = 2$ for carbon and nitrogen) and the potential groups of resources in the diet are more numerous than $n + 1$. Therefore, their use entails methodological decisions about how to aggregate prey with similar isotopic values to reduce the number of sources in the system, while following meaningful ecological criteria to explain human dietary choices (Phillips et al. 2014). To characterize the diet composition

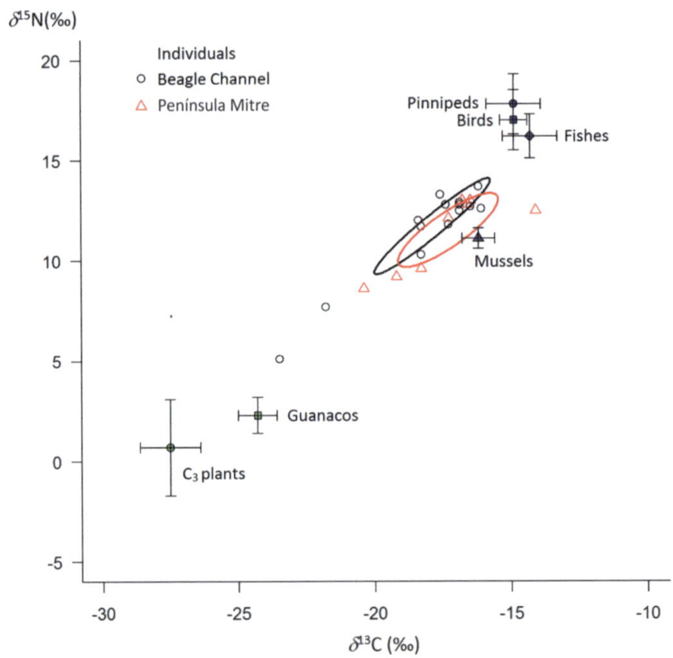

Fig. 5.4 Mixing diagram with main dietary resources and estimated diets for individuals from two geographical sectors: Beagle Channel and Península Mitre. The overlap of the isotopic niches, depicted by corrected standard ellipses, suggests dietary continuities along the southern coast of Tierra del Fuego

of the hunter-gatherers of the southern coast of Tierra del Fuego, we aggregated the resources into four groups: (1) terrestrial plants are the most depleted resource both in $\delta^{13}C$ and $\delta^{15}N$; (2) guanacos are considered separately from plants as the main terrestrial prey; (3) albatrosses are seabirds with the highest $\delta^{15}N$ values and can be interpreted as resources obtained in the pelagic environment (Tivoli 2012); (4) the group of "aggregated marine resources" includes heterogeneous prey: pinnipeds, cormorants, penguins, snoek, merluccid hake, southern hake, and mussels.

Overall, pinnipeds, seabirds, fishes, and mussels contributed between at least one-third and more than half of an individual's diet over the long term (Kochi 2017). Albatrosses were a regular part of the diet, albeit in low frequencies. Tivoli (2012) warns that nowadays species such as black-browed albatrosses can approach the coasts. But she also points out that the increase in the representation of these taxa 1500 years ago coincides with the greater abundance of snoeks in the assemblages, which necessarily implies exploitation of pelagic environments. Indirectly, the isotopic data could also point to the inclusion of offshore environments in the foraging range of hunter-gatherers during the Late Holocene (Tivoli and Zangrando 2011). The intake of plants, in low but isotopically detectable amounts, deserves a mention. They tend to be underestimated in paleodiets from subpolar environments. On the other hand, their exclusion could generate an overestimation of the intake of guanacos, if linear mixing models consider only animal protein (Kochi 2017).

Considering the overlap of different marine taxa in bulk analyses, compound-specific stable isotopes could improve the resolution of paleodietary reconstructions. $\delta^{13}C$ measured in essential amino acids (AA) define patterns or "fingerprints" that discriminate between different primary producer biochemical pathways (Larsen et al. 2013). As such values depend on constant biochemical processes within a taxon, they do not vary seasonally or by locality. Also, "fingerprints" persist along the food chain and are observed in top consumers (Whiteman et al. 2019). With respect to nitrogen, "source" AAs such as phenylalanine (Phe) have a very low discrimination factor by trophic level ($\Delta_{Phe} = 0.4 \pm 0.5‰$, Chikaraishi et al. 2014). Therefore, $\delta^{15}N_{Phe}$ values in consumers represent a weighted average of resources at the base of the trophic chain (Chikaraishi et al. 2014). Trophic amino acids, such as glutamic acid (Glu), fractionate much more with trophic level ($\Delta_{Glu} = 8 \pm 1.2‰$). These differences are related to the AAs metabolism. As isotopic values of AA are measured from a single organism, it is possible to calculate the trophic level even without isotopic values of primary producers. Therefore, this method overcomes an important issue in archaeological studies, when the ancient primary producers are not available for isotopic analyses. Moreover, these baselines are shifting through time and space.

Compound specific isotopic analyses already showed great results with amino acids from pinnipeds bone collagen (Nye 2019; Zangrando et al. 2022). The $\delta^{13}C_{AA}$ values were analyzed with linear discriminant analysis, and the position of pinnipeds was interpreted relative to coastal and pelagic endmembers: giant kelp, Patagonian squid, and migratory Merluccid fishes. The essential AA patterns of pinnipeds change from nearshore diets to offshore diets from the Middle to Late Holocene, in agreement with bulk $\delta^{13}C$ results. Moreover, essential AA patterns of fur seals and sea lions do not overlap, suggesting competitive exclusion and niche separation between species of pinnipeds. $\delta^{15}N_{Phe}$ showed high variability over time, pointing out that a shift in baseline $\delta^{15}N$ is responsible for much of the range in bulk $\delta^{15}N$ of consumers (Nye 2019). In this case, humans seemed to have little impact on the trophic level of pinnipeds over time (Saporiti et al. 2014; Vales et al. 2017).

As a summary of this second set of isotopic studies on the southern coast of Tierra del Fuego, we highlight the focus on marine and terrestrial food webs as a whole system. In a recent work, Szpak (2022: 260) answered why zooarchaeology needs stable isotope analyses, as this technique tells us about how animals actually *lived* and interacted with people previous to their capture, a time span (long) before their death. The characterization of prey ecological niches through intrinsic markers in their tissues definitely enriched our knowledge about human-animal interactions, a lot more about those relations below and above the water surface, near the shores, and offshore habitats. We reconsidered the impact of human exploitation, which seems to be more profound in pinniped populations than fishes or guanacos in the forest. However, the analyses of food web structures are still snapshots of particular moments of the Holocene. A different "film" may emerge analyzing more episodes related to climatic fluctuations, and using other tools such as compound-specific stable isotopes analyses. The isotopic characterization of different taxa defined the scopes and limitations of bulk analyses, particularly in humans as omnivores. Bone collagen isotopic analyses do not have enough resolution to evaluate offshore-nearshore gradients in the average human diet. Instead, several lines of evidence need to be integrated to evaluate adaptive trajectories, the logistics of seafaring, and the lifestyles by the sea (Bjerck and Zangrando 2016).

5.4 A Challenging Environment and Changing Times: Diets, Nutrition, and Health

Paleodietary reconstructions with stable isotopes can also contribute to the health-related aspects of nutrition. In this perspective, it is considered the macronutrients and micronutrients supplied by food, whose deficiency or excess have effects on the health of individuals and impact demographic trends in populations (Hockett and Haws 2003). The most known application of stable isotopes in these topics is the characterization of macronutrient intake, specifically the energy-containing substances such as carbohydrates, lipids, and proteins (Ulijaszek et al. 2012). For example, the use of $\delta^{15}N$ values to assess the relative proportion of proteins in the diet. Low $\delta^{15}N$ values in a tissue, when combined with skeletal or dental indicators of undernutrition, may aid the differential diagnoses of ill health (Reitsema and Holder 2018).

Tierra del Fuego, as a high-latitude island environment, exhibits a strongly seasonal distribution of resources, with the majority of accessible energy derived from secondary biomass. Therefore, populations are more dependent on animal prey than on plants for subsistence (Cordain et al. 2000; Binford 2001). The ecological setting and $\delta^{15}N$ values in humans, both from the southern coast and the northern steppe, point to diets relatively rich in animal protein of terrestrial and/or marine origin. Schinder and Guichón (2003) observed that porotic hyperostosis in these individuals would not have its origin in chronic dietary deficiencies. Nutrient loss has been proposed as one of the causes, from exposure to parasites from raw fish and marine mammal meat (Aspillaga et al. 1999); alternately, other dietary factors may also inhibit iron absorption (Suby 2014b). Other non-specific indicators of physiological stress were observed in very low frequencies, such as linear enamel hypoplasias (Schinder and Guichón 2003; Tessone et al. 2011).

Patterns of physical activity were analyzed more broadly with $\delta^{15}N$ values, with diets being classified either as marine, mixed, or terrestrial. The presence of osteoarticular disorders and enthesopathies is in agreement with some individuals with marine diets. Frequent injuries in the shoulder region, followed by the rest of the upper limb, are associated with paddling (Aspillaga et al. 2006; Suby et al. 2011; Tessone et al. 2011). Some of the indicators examined in relation to diet type were the femoral robusticity index (Suby and Guichón 2009), pathological joint diseases such as the osteochondritis dissecans (OCD; Zúñiga Thayer et al. 2020) and Schmorl's nodes (SN; Suby 2014a). The prevalence of OCD did not show a significant difference between individuals with a terrestrial, mixed, or marine diet in the larger sample of southern Patagonia. On the other hand, the prevalence of SN seems to follow a geographical pattern, because it is more frequent in continental Patagonia than Tierra del Fuego as a whole. Individuals with marine diets did not present SN, but small sample size bias calls for caution in these interpretations. In addition to the impact of physical activities associated with economic strategies, genetically influenced developmental defects may also contribute to the prevalence of joint diseases (Suby 2014b; Zúñiga Thayer et al. 2020).

Other works characterized only the relative macronutrient composition of diets as a complement of prey categories, in the framework of optimal foraging models (Schinder et al. 2003). One way to assess the protein and non-protein components of the diet is with the difference between the $\delta^{13}C_{apatite}$ and $\delta^{13}C_{collagen}$ values in the same individual, which is $\Delta^{13}C_{apa\text{-}col}$ (Ambrose et al. 1997). Monoisotopic diets include both protein and energy (lipids and carbohydrates) from C_3 sources. It is estimated that individuals consuming these diets present $\Delta^{13}C_{apa\text{-}col}$ values around 4.4‰. This difference varies if terrestrial and marine resources are combined. On the southern coast of Tierra del Fuego, $\Delta^{13}C_{apa\text{-}col}$ analysis suggested that 11 individuals from the northern coast of Peninsula Mitre ($n = 6$) and Beagle Channel ($n = 5$) had a high consumption of protein and lipids, attributed to blubber of marine mammals (Schinder et al. 2003). However, at larger spatial scales, cluster analysis indicates that carbohydrate and lipid consumption is not solely defined by geographic sector or diet type (Barberena 2004).

Reviews of experimental and field studies caution that absolute $\Delta^{13}C_{apa\text{-}col}$ values alone do not diagnose a particular protein or total diet combination (Kellner and Schoeninger 2007). This methodology also depends on the most problematic link between isotope theory and its use in ecology, which is the measurement and choice of discrimination factors (Bond and Diamond 2011). An alternative is a model based on the multivariate analysis of $\delta^{13}C_{col}$, $\delta^{13}C_{apa}$, and $\delta^{15}N$ values (Froehle et al. 2012). There are five clusters defined from the isotopic measurements on 158 individuals of eight prehistoric populations. Their dietary compositions are derived from the comparison with animal isotopic data under controlled diets, as well as from other lines of evidence (Kellner and Schoeninger 2007). This database generates discriminant functions that classify new individuals into the five groups characterized as diets with (1) 100% C_3 protein, 100% C_3 energy; (2) > 50% C_4 protein, 30% C_3 and 70% C_4 energy; (3) marine protein, 50% C_3 and 50% C_4 energy; (4) $\geq$ 65% C_3 protein, 70% C_3 and 30% C_4 energy; (5) $\geq$ 65% C_3 protein, 30% C_3 and 70% C_4 energy (Froehle et al. 2012). In high-latitude marine environments, the greatest potential of the multivariate model would be the ability to distinguish quantitative input between C_3 and C_4 energy; the latter are interpreted as corresponding to marine fats, because they tend to higher $\delta^{13}C$ values than those of terrestrial animals (Heron et al. 2013).

In the article by Froehle et al. (2012), the individuals from the north of Tierra del Fuego, the southern coast and north coast of the Península Mitre ($n = 11$) are distributed between cluster 3 and cluster 1. Unlike the analyses based on $\Delta^{13}C_{apa\text{-}col}$, they suggest an important contribution of C_3 energy. In a posterior work (Alunni et al. 2021) we reanalyzed these discrepancies, combining stable isotopes with zooarchaeological studies. The human sample was expanded with 14 individuals from the southern coast, and the discussion integrated the butchering and bone processing techniques related to the preparation of guanaco bone grease.

Most of the new individuals were also classified in cluster 3 of the multivariate model: a diet of marine protein, 50% C_3 and 50% C_4 energy, the latter component interpreted as corresponding to fats of marine origin (Fig. 5.5). The proportion of C_3 energy, such as the plants and fat from terrestrial animals, seemed relatively high in a predominantly marine diet. However, in some zooarchaeological assemblages the most abundant prey is the guanaco. In particular localities it is observed a dominance of bones with high marrow content (e.g., upper, middle and lower limbs, calcaneus, and phalanges) over the axial region elements with high meat yield. In addition, bones with high marrow content show intense processing, such as the high percentage of fresh fractures and negative impacts, the low proportion of complete bone remains, and morphological criteria in epiphyseal fragments related to boiling (Morin and Soulier 2017). These processing patterns are not found in all coastal sectors, as the zooarchaeological assemblages could be reflecting the situational processing of guanaco fats related to their spatiotemporal availability. This contrasts with the estimated uniformity of the energetic portion of the diet through stable isotopes. Therefore, it is suggested that hunter-gatherers practiced certain degree of storage and scheduled handling of fats to secure its provision in the area and throughout the annual cycle. One of the individuals could not be assigned to any cluster, such as the female adult from the Acatushún Cemetery, Harberton Ranch. This divergence may be indicative of post-contact dietary changes, possibly linked to the introduction of European food sources or altered subsistence practices in the colonial context.

The intensity of the contact between the native Yaghan populations and the Europeans increased dramatically at the mid-end nineteenth century, with the scientific/military expeditions (Fiore and Varela 2009), the permanent establishment of the

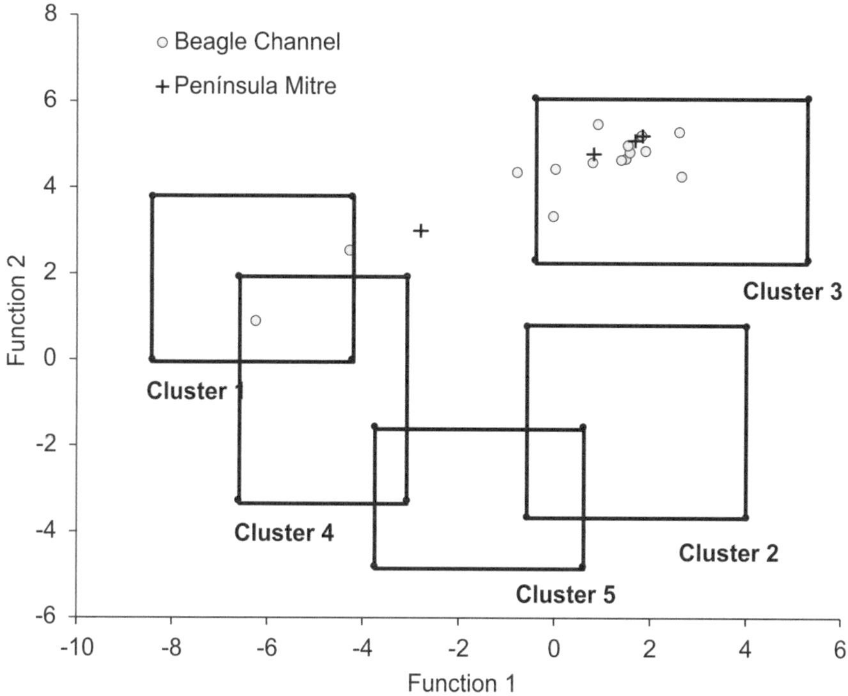

Fig. 5.5 Individuals from the Beagle Channel and Península Mitre, classified in clusters with diets characterized by Froehle et al. (2012). See references in Alunni et al. (2021)

Anglican missions (Martinic Beros 2001), the process of formal occupation of the southern region by the Argentine and Chilean governments, and finally, the imposition of a new socio-economic structure (Bitlloch 2005). Even in this context, the sources record an oscillatory pattern between sedentarism and nomadism (Spielmann and Eder 1994), which has also been observed in other cases (Valenzuela et al. 2019). Tafuri et al. (2017) addressed the transculturation process in southern Tierra del Fuego, through the $\delta^{13}C_{col}$ and $\delta^{15}N$ measurements on 25 individuals from the nineteenth century (Marangoni et al. 2011). The bioanthropological collection was assembled by the Italian captain Giacome Bove, during his first (1881–1882) and second expeditions (1883–1884) to Tierra del Fuego. The exact provenance and chronology of the second collection, housed at the Museum of Anthropology in Rome, is more difficult to contextualize. Despite these problems, it is observed that both samples come from populations with similar distributions of $\delta^{15}N$ values. The average $\delta^{15}N$ tends to be more positive in the postcontact sample than in the precontact sample, being $+18.4 \pm 1.5‰$ and $+17.2 \pm 2.5‰$ respectively. Statistically there is no difference between the $\delta^{13}C$ averages: $-12.2 \pm 0.8‰$ (postcontact) and $-12.8 \pm 2.1‰$ (precontact). These data were interpreted in terms of cultural resilience, in that pinnipeds were supplemented by other high trophic level marine resources to maintain protein intakes (Tafuri et al. 2017).

The marine hunter-gatherer individuals studied by Tafuri et al. (2017) probably still reflect an early phase of colonization, and the energy portion of the diets remains uncharacterized without isotopic analyses on carbonate. Valenzuela et al. (2019) reported that members of the Selk'nam community, buried in the Salesian mission "La Candelaria" during 1897–1948, did undergo a pronounced dietary change. This study corresponds to a later phase of colonization. The group from the cemetery has significantly lower values in bone collagen $\delta^{13}C$ and $\delta^{15}N$ compared with the precontact sample from the northern steppe. Intraindividual isotopic variation, comparing teeth and bone, shows a similar trend in the postcontact sample. Historical diets in the mission were described as unbalanced, with a higher proportion of carbohydrates and a lower proportion of animal proteins. However, it is possible that the historical dietary pattern, both at the north and south of Tierra del Fuego, reflects not only the different chronological phases of the colonization, but also different responses to the process. Several written sources about the Yaghan mention an oscillatory pattern between sedentarism and nomadism (Gusinde 1986 [1937]), which seems to have concerned the Anglican missionaries (Bove 2005 [1883]). The traditional subsistence systems may have been reorganized in face to the colonial context, but the access to marine resources could have complemented the unequal distribution of wealth, and navigation may facilitate avoidance strategies (Rankin and Crompton 2016). The terrestrial hunter-gatherers may be more critically affected by the cattle ranching economy, the constrained mobility and territorial disintegration (Borrero 1994). Coastal native populations were quickly disrupted by the onset of maritime exploration by European nations (Bailey and Milner 2002), but further work with stable isotopes may deepen in the extension and complexities of contact, adaptation and transculturation at the end of the world.

In summary, a body of research conducted in Tierra del Fuego can be identified that addresses—either explicitly or implicitly—the relationship between diet, nutrition, and health through stable isotope analyses. The $\delta^{15}N$ was the main proxy to evaluate protein intake, associated with non-specific indicators of stress. Common indicators such as porotic hyperostosis and cribra orbitalia in this group of marine hunter-gatherers may be related to parasite infections or reduced iron absorption, because high $\delta^{15}N$ values do not suggest chronic dietary deficiencies. Indicators of physical activity in some individuals, especially those in the upper limbs, are coherent with marine diets obtained by seafaring logistics. On the other hand, pathological joint diseases do not follow an evident pattern by diet type. As the recording and interpretation of lesions related to infections are increasingly systematized in southern Patagonia (Pandiani et al. 2022), future works could deepen in the relationship between types of pathological conditions and stable isotope ratios in skeletal tissues (Reitsema and Holder 2018). Moreover, the integration of paleopathological approaches with stable isotope analyses has an enormous potential to assess the impact of colonization from a biocultural perspective (Goodman et al. 1988): for example, hypothesizing changes in access to food resources, situations of inequality, and deteriorated health. For now, bioarchaeological studies of individuals from the southern coast of Tierra del Fuego did not find significant changes between pre and postcontact periods, although the sample size is still small.

We also reviewed stable isotope studies focusing on the consumption of macronutrients, specifically the relative intake of marine/terrestrial protein and energy (Schinder et al. 2003; Alunni et al. 2021). In high-latitude environments, a closer look on the fat intake helps to understand the dietary choices as part of the adaptive strategies to their environment (Haas and Harrison 1977). The interpretive models for isotopic measurements (Ambrose et al. 1997; Froehle et al. 2012) may be subjected to new experimental findings, revisions and expanding human databases (Kellner and Schoeninger 2007; Zhu and Sealy 2019). Meanwhile, we can overcome these limitations relying on different lines of evidence. Here we discussed the herbivore bone processing strategies for fat extraction, supporting the substantial contribution of C_3 energy in a predominantly marine diet. In continental Patagonia, additional studies have evaluated technological strategies for the processing, cooking, and storage of animal fat, based on the isotopic analysis ($\delta^{13}C$ and $\delta^{15}N$) of organic residues recovered from archaeological contexts (Chaile et al. 2018). The isotopic record contributes to characterize not only the consumption of prey groups or types, but also the previous phases of the procurement, distribution, preparation and disposal of foods.

5.5 Conclusion: Thirty Years of Stable Isotopes Analysis in Tierra del Fuego

Archaeological projects in coastal Patagonia explored the historical trajectories of native populations in Patagonia, demonstrating the variability in the exploitation of aquatic resources and the occupation of coastal areas, in intensity and frequency, throughout the Holocene (Borrero and Barberena 2006; Orquera and Gómez Otero 2007). The subsistence system of specialized marine hunter-gatherers poses a particular challenge: how to connect the activities carried out on the coasts, on the vessels floating offshore and inland as far as the waters are no longer visible? (Ford 2011). The stable isotope studies in Tierra del Fuego contributed greatly to this topic. They assessed not only the intensity of marine resource consumption, which would be a "classic" approach, but they also discussed the logistics of seafaring in an inshore-offshore gradient and the management of terrestrial fats, highlighting the flexibility of subsistence strategies in a Sub-Antarctic environment.

The assessment of subsistence strategies was related to an even more challenging subject, which is the biocultural interaction between the different hunter-gatherer groups inhabiting the southern coast of Tierra del Fuego during the Late Holocene (Vázquez et al. 2011). These questions are among the classic topics of hunter-gatherer archaeology (Binford 2002), but the answers were strongly directed by written sources during the last four centuries (Borrero 1997; 2011). For these questions, stable isotopes offered insights about the cultural practices connected to diet and mobility to infer intra- and intergroup interactions. We emphasize once again that the characteristics of the local isotopic ecology determine the scope and limitations of subsequent interpretations of human isotopic data. Carbon and nitrogen isotopic composition of dietary resources from the southern coast of Tierra del Fuego have significant variations at different spatial and taxonomic levels, but in the long-term human diets it is inferred a paleodietary homogeneity for the entire southern coast of Tierra del Fuego (Tessone et al. 2011), including the Beagle Channel and the southern coast of Península Mitre. Most individuals have predominantly marine diets complemented with terrestrial resources, supporting a long-term, continuous bond between human groups and the marine environment.

This strong linkage of humans as consumers at multiple levels of the local food web (Maschner et al. 2009) lead us to reflect about the anthropic impact on the environment. Bulk and compound specific stable isotope analyses suggest that even a small-scale society could have exerted considerable predatory pressure over large animal prey, such as the fur seals and sea lions. The historical ecology of pinnipeds and other marine fauna, pondering the effects of climatic change and anthropic impacts, is under discussion over the Patagonian coast (Zenteno et al. 2015; Bas et al. 2019). Different approaches may contribute to this discussion, but isotopic studies must work on a temporally continuous baseline to correctly assess the changes in habitat or trophic position.

The exploration of dietary patterns through time and in relation with sex and age categories is a pending matter, also considering the small sample sizes and that in several cases the human remains were recovered without archaeological methods. However, stable isotopes are already discussing the effects of the colonization processes in Tierra del Fuego, one of the latest areas incorporated to the modern states of Argentina and Chile. Ethnographic sources indicate that most of the foods introduced to the native diets were carbohydrates: those derived from flour and sugar. Therefore, $\delta^{13}C$ measurements on bioapatite may reveal a different rate of dietary change compared to isotopic analyses on collagen, as the latter one is synthesized by the preferential routing of proteins. For now, the $\delta^{13}C_{col}$ and $\delta^{15}N$ data on the historical individuals suggest that several aspects of the marine lifestyle may have persisted into the beginning of the twentieth century, despite the crisis brought by the epidemics and the imposition of a new socioeconomic order.

References

Albero MC, Angiolini FE, Piana EL (1986) Discordant ages related to reservoir effect of associated archaeologic remains from the Tunel Site, Beagle Channel, Argentine Republic. Radiocarbon 28(2A):748–753. https://doi.org/10.1017/S0033822200007980

Alfonso-Durruty MP, Giles BT, Misarti N, San Román M, Morello F (2015) Antiquity and geographic distribution of cranial modification among the prehistoric groups of Fuego-Patagonia, Chile. Am J Phys Anthropol 158(4):607–623. https://doi.org/10.1002/ajpa.22832

Alunni DV, Kochi S, Tessone A, Zangrando AFJ (2021) Consumo de grasas entre cazadores-recolectores de la costa sur de Tierra del Fuego durante el Holoceno tardío: una evaluación desde la zooarqueología y los isótopos estables. Comechingonia 25(3):61–70. https://doi.org/10.37603/2250.7728.v25.n3.32419

Ambrose SH, Butler BM, Hanson DB, Hunter-Anderson RL, Krueger HW (1997) Stable isotopic analysis of human diet in the Marianas Archipelago, Western Pacific. Am J Phys Anthropol 104(3):343–361. https://doi.org/10.1002/(SICI)1096-8644(199711)104:3

Arneborg J, Heinemeier J, Lynnerup N, Nielsen HL, Rud N, Sveinbjörnsdóttir ÁE (1999) Change of diet of the Greenland Vikings determined from stable carbon isotope analysis and ^{14}C dating of their bones. Radiocarbon 41(2):157–168. https://doi.org/10.1017/S0033822200019512

Aspillaga E, Ocampo C, Rivas P (1999) Restos óseos humanos de contextos arqueológicos del área de isla Navarino: indicadores de estilos de vida en indígenas canoeros. Anal Inst Patag 27:123–135

Aspillaga E, Castro M, Rodriguez M, Ocampo C (2006) Paleopatología y estilo de vida: El ejemplo de los Chonos. Magallania 34(1):77–85. https://doi.org/10.4067/S0718-22442006000100005

Bailey G, Milner N (2002) Coastal hunter-gatherers and social evolution. Before Farming 2002(3–4):1–23. https://doi.org/10.3828/bfarm.2002.3-4.1

Barberena R (2004) Arqueología e isótopos estables en Tierra del Fuego. In: Borrero LA, Barberena R (eds) Temas de Arqueología. Arqueología del norte de la Isla Grande de Tierra del Fuego. Editorial Dunken, Buenos Aires, pp 132–169

Barberena R, Tessone A (2018) Isotopic studies of foragers' diet: environmental archaeological approaches. In: Smith C (ed) Encyclopedia of global archaeology. Springer, Cham, pp 1–11

Barberena R, L'Heureux GL, Borrero LA (2004) Expandiendo el alcance de las reconstrucciones de subsistencia. Isótopos estables y conjuntos arqueofaunísticos. In: Civalero MT, Fernández PM, Guráieb AG (eds) Contra viento y marea. Arqueología de Patagonia. INAPL, Sociedad Argentina de Antropología, Buenos Aires, pp 417–433

Bas M, Briz i Godino I, Álvarez MR, Vales DG, Crespo EA, Cardona L (2019) Back to the future? Late Holocene marine food web structure in a warm climatic phase as a predictor of trophodynamics in a warmer South-Western Atlantic Ocean. Glob Change Biol 25(2):404–419. https://doi.org/10.1111/gcb.14523

Binford LR (2001) Constructing frames of reference: an analytical method for archaeological theory building using hunter-gatherer and environmental data sets. University of California Press, Berkeley

Binford LR (2002) In pursuit of the past: decoding the archaeological record. University of California Press, Berkeley and Los Angeles

Bitlloch RE (2005) Antiguos pueblos del extremo sur del mundo. Libros del Huerquén, Buenos Aires

Bjerck HB, Zangrando AFJ (2016) Introduction: marine ventures—archaeological perspectives on human-sea relations. In: Bjerck HB, Breivik HM, Fretheim SE, Piana E, Skar B, Tivoli AM, Zangrando AFJ (eds) Marine ventures—archaeological perspectives on human-sea relations. Equinox, Sheffield, pp 3–14

Bond AL, Diamond AW (2011) Recent Bayesian stable-isotope mixing models are highly sensitive to variation in discrimination factors. Ecol Appl 21(4):1017–1023. https://doi.org/10.1890/09-2409.1

Borrero LA (1994) The extermination of the Selk'nam. In: Ellanna LJ, ES Burch (eds) Key issues in hunter-gatherer research. Routledge, New York, pp 247–261

Borrero LA (1997) The origins of ethnographic subsistence patterns in Fuego-Patagonia. In: McEwan C, Borrero LA, Prieto A (eds) Patagonia. Natural history, prehistory and ethnography at the uttermost end of the earth. Princeton University Press, New Jersey, pp 60–81

Borrero LA (2011) Los cazadores orientales de Tierra del Fuego. In: Zangrando AFJ, Vázquez M, Tessone A (eds) Los Cazadores-Recolectores del Extremo Oriental Fueguino. Arqueología de Península Mitre e Isla de los Estados. Sociedad Argentina de Antropología, Buenos Aires, pp 287–298

Borrero LA, Barberena R (2006) Hunter-gatherer home ranges and marine resources: an archaeological case from southern Patagonia. Curr Anthropol 47(5):855–867. https://doi.org/10.1086/507186

Borrero LA, Guichón RA, Tykot R, Kelly J, Prieto A, Cárdenas P (2001) Dieta a partir de isótopos estables en restos óseos humanos de Patagonia Austral. Estado actual y perspectivas. Anal Inst Patag Ser Cs Hum 29:119–127

Bove G [1883] (2005) Expedición a la Patagonia: Un viaje a las tierras y mares australes (1881–1882). Ediciones Continente, Buenos Aires

Casey MM, Post DM (2011) The problem of isotopic baseline: reconstructing the diet and trophic position of fossil animals. Earth Sci Rev 106:131–148. https://doi.org/10.1016/j.earscirev.2011.02.001

Chaile C, Kochi S, Pérez SA, Ceballos L (2018) Cazar y recolectar: Aportes interdisciplinarios para pensar la nutrición en estudios arqueológicos. Zaranda Ideas 16(2):44–60

Chapman A, Hester TR (1973) New data on the archaeology of the Haush: Tierra del Fuego. J Soc Américanistes LXII:185–208. https://www.jstor.org/stable/24605455

Chikaraishi Y, Steffan SA, Ogawa NO, Ishikawa NF, Sasaki Y, Tsuchiya M, Ohkouchi N (2014) High-resolution food webs based on nitrogen isotopic composition of amino acids. Ecol Evol 4(12):2423–2449. https://doi.org/10.1002/ece3.1103

Chisholm BS, Nelson DE, Schwarcz HP (1982) Stable-carbon isotope ratios as a measure of marine versus terrestrial protein in ancient diets. Science 216(4550):1131–1132. https://doi.org/10.1126/science.216.4550.1131

Clementz MT, Koch PL (2001) Differentiating aquatic mammal habitat and foraging ecology with stable isotopes in tooth enamel. Oecologia 129:461–472. https://doi.org/10.1007/s004420100745

Coiazzi A (1914) Los indios del archipiélago fueguino. Rev Chil Hist Geogr 13:1–113

Cordain L, Miller JB, Eaton SB, Mann N, Holt SH, Speth JD (2000) Plant-animal subsistence ratios and macronutrient energy estimations in worldwide hunter-gatherer diets. Am J Clin Nutr 71(3):682–692. https://doi.org/10.1093/ajcn/71.3.682

Fiore D, Varela ML (2009) Memorias de papel: Una arqueología visual de las fotografías de pueblos originarios fueguinos. Dunken, Buenos Aires

Fitz Roy R (1839) Proceedings of the second expedition, 1826–1830, under the command of Captain P. Parker King, R. N., F. R. S. In: Narrative of the surveying voyages of his majesty's ships adventure and Beagle, between the years 1826–36, describing their examination of the southern shores of South America, and the Beagle circumnavigation of the globe, vol 1. Henry Colburn, London

Ford B (2011) Coastal archaeology. In: Catsambis A, Ford B, Hamilton DL (eds) The Oxford handbook of Maritime archaeology. Oxford University Press, New York, pp 763–785

France R (1995) Carbon-13 enrichment in benthic compared to planktonic algae: food web implications. Mar Ecol Prog Ser 124(1–3):307–312. https://doi.org/10.3354/meps124307

Froehle AW, Kellner CM, Schoeninger MJ (2010) FOCUS: effect of diet and protein source on carbon stable isotope ratios in collagen: follow up to Warinner and Tuross (2009). J Archaeol Sci 37(10):2662–2670

Froehle AW, Kellner CM, Schoeninger MJ (2012) Multivariate carbon and nitrogen stable isotope model for the reconstruction of prehistoric human diet. Am J Phys Anthropol 147(3):352–369. https://doi.org/10.1002/ajpa.21651

FUCOA (2014) Kawésqar: serie introducción histórica y relatos de los pueblos de Chile. Fundación de Comunicaciones, Capacitación y Cultura del Agro, Santiago de Chile

Furlong CW (1917) Tribal distribution and settlements of the fuegians, comprising nomenclature, etymology, philology, and populations. Geogr Rev 3(3):169–187. https://doi.org/10.2307/207659

Goodman AH, Brooke Thomas R, Swedlund AC, Armelagos GJ (1988) Biocultural perspectives on stress in prehistoric, historical, and contemporary population research. Am J Phys Anthropol 31:169–202. https://doi.org/10.1002/ajpa.1330310509

Goody J (1982) Cooking, cuisine and class: a study in comparative sociology. Cambridge University Press, Cambridge

Guichón RA, Borrero LA, Prieto AI, Cárdenas P, Tykot R (2001) Nuevas determinaciones de isótopos estables Para Tierra del Fuego. Rev Argent Antropol Biol 3(1):113–126. http://sedici.unlp.edu.ar/handle/10915/5621

Gusinde M [1937] (1986) Los indios de Tierra del Fuego, tomo segundo, vol I: Los Yámana. Centro Argentino de Etnología Americana, Buenos Aires

Haas JD, Harrison GG (1977) Nutritional anthropology and biological adaptation. Annu Rev Anthropol 6(1):69–101. https://www.jstor.org/stable/2949326

Hedges RE, Reynard LM (2007) Nitrogen isotopes and the trophic level of humans in archaeology. J Archaeol Sci 34(8):1240–1251. https://doi.org/10.1016/j.jas.2006.10.015

Heron C, Andersen S, Fischer A, Glykou A, Hartz S, Saul H, Steele V, Craig O (2013) Illuminating the Late Mesolithic: residue analysis of 'blubber' lamps from Northern Europe. Antiquity 87(335):178–188. https://doi.org/10.1017/S0003598X00048705

Hobson KA (2007) Isotopic tracking of migrant wildlife. In: Michener R, Lajtha K (eds) Stable isotopes in ecology and environmental science. Wiley, Singapur, pp 155–170

Hockett B, Haws J (2003) Nutritional ecology and diachronic trends in Paleolithic diet and health. Evol Anthropol 12:211–216. https://doi.org/10.1002/evan.10116

Hyades PD, Deniker J (1891) Anthropologie et Ethnographie. Mission Scientifique du Cap Horn (1882–1883), vol VII. Ed. oficial, Paris

Jackson AL, Parnel LAC, Inger R, Bearhop S (2011) Comparing isotopic niche widths among and within communities: SIBER—stable isotope Bayesian ellipses in R. J Anim Ecol 80:595–602

Kellner CM, Schoeninger MJ (2007) A simple carbon isotope model for reconstructing prehistoric human diet. Am J Phys Anthropol 133(4):1112–1127. https://doi.org/10.1002/ajpa.20618

Kochi S (2017) Paleodietas en cazadores-recolectores del canal Beagle durante el Holoceno tardío. Intersecc Antropol 18(3):329–339

Kochi S, Pérez SA, Tessone A, Ugan A, Tafuri MA, Nye J, Tivoli AM, Zangrando AFJ (2018) $\delta^{13}C$ and $\delta^{15}N$ variations in terrestrial and marine food webs of Beagle Channel in the Holocene. Implications for human paleodietary reconstructions. J Archaeol Sci Rep 18:696–707. https://doi.org/10.1016/j.jasrep.2017.11.036

Kochi S, Tessone A, Zangrando AF, Ugan A, Alunni D, Salemme M, Santiago F, Vázquez M (2020) $\delta^{13}C$ and $\delta^{15}N$ variation on guanacos in Tierra del Fuego: references for terrestrial food webs in a high-latitude Island. Archaeol Anthropol Sci 12:52. https://doi.org/10.1007/s12520-020-01011-7

Larsen T, Ventura M, Andersen N, O'Brien DM, Piatkowski U, McCarthy MD (2013) Tracing carbon sources through aquatic and terrestrial food webs using amino acid stable isotope fingerprinting. PLoS One 8(9):e73441. https://doi.org/10.1371/annotation/2f7a422e-13a9-4491-ae81-4fd215149654

Layman CA, Araujo MS, Boucek R, Hammerschlag-Peyer CM, Harrison E, Jud ZR, Matich P, Rosenblatt AE, Vaudo JJ, Yeager LA, Post DM, Bearhop S (2012). Applying stable isotopes to examine food-web structure: an overview of analytical tools. Biol Rev 87(3):545–562. https://doi.org/10.1111/j.1469-185X.2011.00208.x

Leng MJ, Lewis JP (2016) Oxygen isotopes in Molluscan shell: applications in environmental archaeology. Environ Archaeol 21(3):295–306. https://doi.org/10.1179/1749631414Y.0000000048

Lothrop SK (2022 [1928]) The Indians of Tierra del Fuego: an account of the Ona, Yahgan, Alacaluf and Haush natives of the Fuegian Archipelago. Zagier & Urruty Publications, Ushuaia

Marangoni A, Belli ML, Caramelli D, Moggi Cecchi J, Zavattaro M, Manzi G (2011) Tierra del Fuego, its ancient inhabitants, and the collections of skeletal remains in the museums of anthropology of Florence and Rome. Museol Sci 5(1–2):88–96

Martinic Beros M (2001) El postrer esfuerzo misional entre los yámana (1888–1917) significación en la decadencia étnica. Estado de la comunidad final (1918–2000). Anal Inst Patag Ser Cs Hum 29:5–27

Martinoli MP (2018) Modalidades de explotación, procesamiento y consumo de pinnípedos en la margen meridional de Tierra del Fuego. Ph.D. dissertation, Universidad de Buenos Aires

Maschner HD, Betts MW, Cornell J, Dunne JA, Finney B, Huntly N, Benson B (2009) An introduction to the biocomplexity of Sanak Island, Western Gulf of Alaska. Pac Sci 63(4):673–709. https://doi.org/10.2984/049.063.0410

Minagawa M, Wada E (1984) Stepwise enrichment of ^{15}N along food chains: further evidence and the relation between δ^{15}N and animal age. Geochim Cosmochim Acta 48(5):1135–1140. https://doi.org/10.1016/0016-7037(84)90204-7

Misarti N, Finney B, Maschner H, Wooller MJ (2009) Changes in northeast Pacific marine ecosystems over the last 4500 years: evidence from stable isotope analysis of bone collagen from archeological middens. The Holocene 19(8):1139–1151. https://doi.org/10.1177/0959683609345075

Morin E, Soulier MC (2017) New criteria for the archaeological identification of bone grease processing. Am Antiq 82(1):96–122. https://doi.org/10.1017/aaq.2016.16

Newsome SD, Martínez del Río C, Bearhop S, Phillips DL (2007) A niche for isotopic ecology. Front Ecol Environ 5:429–436. https://doi.org/10.1890/060150.1

Nye JW (2019) Prehistoric and modern ecological dynamics in southern South American marine food webs. Ph.D. dissertation, University of California, Riverside

Nye JW, Zangrando AFJ, Martinoli MP, Fogel ML (2020) Temporal and population trends in human exploited pinnipeds from Tierra del Fuego. Palaeogeogr Palaeoclimatol Palaeoecol 554:109804. https://doi.org/10.1016/j.palaeo.2020.109804

Orquera LA, Gómez Otero J (2007) Los cazadores-recolectores de las costas de Pampa, Patagonia y Tierra del Fuego. Rel Soc Arg Antropol (32):75–100

Orquera LA, Piana LE (1996) El sitio Shamakush I (Tierra del Fuego, República Argentina). Rel Soc Arg Antropol XXI:215–265

Orquera LA, Piana LE (1999a) Arqueología de la región del canal Beagle (Tierra del Fuego, República Argentina). Sociedad Argentina de Antropología, Buenos Aires

Orquera LA, Piana LE (1999b) La vida material y social de los Yámana. EUDEBA, Buenos Aires

Orquera LA, Piana LE (2009) Sea nomads of the Beagle channel in southernmost South America: over six thousand years of coastal adaptation and stability. J Isl Coast Archaeol 4:61–81. https://doi.org/10.1080/15564890902789882

Panarello HO (1987) Oxygen-18 temperatures on present and fossil invertebrated shells from Túnel Site, Beagle Channel, Argentina. In: Rabassa J (ed) Quaternary of South America and Antarctic Peninsula. CRC Press, London, pp 83–91

Panarello HO, Zangrando AFJ, Tessone A, Kozameh LF, Testa N (2006) Análisis comparativo de paleodietas humanas entre la región del canal Beagle y Península Mitre: perspectivas desde los isótopos estables. Magallania 34(2):37–46. https://doi.org/10.4067/S0718-22442006000200004

Pandiani CD, Santos AL, Suby JA (2022) Procesos infecciosos en cazadores-recolectores de Patagonia Austral: Evidencias en restos óseos del Holoceno medio y tardío (ca. 5200–100 años aP). Lat Am Antiq 33(2):376–394. https://doi.org/10.1017/laq.2021.64

Phillips DL, Inger R, Bearhop S, Jackson AL, Moore JW, Parnell AC, Semmens BX, Ward EJ (2014) Best practices for use of stable isotope mixing models in food-web studies. Can J Zool 92(10):823–835. https://doi.org/10.1139/cjz-2014-0127

Rankin L, Crompton A (2016) Kayaks and Chaloupes: Labrador Inuit and the seascapes of intercultural contact. In: Bjerck HB, Breivik HM, Fretheim SE, Piana E, Skar B, Tivoli AM, Zangrando AFJ (eds) Marine ventures—archaeological perspectives on human-sea relations. Equinox, Sheffield, pp 383–398

Reitsema LJ, Holder S (2018) Stable isotope analysis and the study of human stress, disease, and nutrition. Bioarchaeol Int 2(2):63–74. https://doi.org/10.5744/bi.2018.1018

Riccialdelli L, Newsome SD, Fogel ML, Goodall RNP (2010) Isotopic assessment of prey and habitat preferences of a cetacean community in the southwestern South Atlantic Ocean. Mar Ecol Prog Ser 418:235–248. https://doi.org/10.3354/meps08826

Riccialdelli L, Newsome SD, Fogel ML, Fernández DA (2017) Trophic interactions and food web structure of a subantarctic marine food web in the Beagle Channel: Bahía Lapataia, Argentina. Pol Biol 40:807–821. https://doi.org/10.1007/s00300-016-2007-x

Richards MP, Hedges REM (1999) Stable isotope evidence for similarities in the types of marine foods used by late Mesolithic humans at sites along the Atlantic Coast of Europe. J Archaeol Sci 26(6):717–722. https://doi.org/10.1006/jasc.1998.0387

Saletta MJ (2017) Los Selk'nam y los Haush de Tierra del Fuego. In: Vázquez M, Elkin D, Oría J (eds) Patrimonio a orillas del mar. Arqueología del Litoral Atlántico de Tierra del Fuego. Editora Cultural de Tierra del Fuego, Ushuaia, pp 103–118

San Román M (2010) La explotación de recursos faunísticos en el sitio Punta Santa Ana 1: estrategias de subsistencia de grupos de cazadores marinos tempranos de Patagonia meridional. Magallania 38(1):183–198. https://doi.org/10.4067/S0718-22442010000100011

Santiago F, Salemme M, Suby J, Guichón RA (2011) Restos humanos en el Norte de Tierra del Fuego: aspectos contextuales, dietarios y paleopatológicos. Intersecc Antropol 12(1):147–162

Saporiti F, Bala LO, Otero JG, Crespo EA, Piana EL, Aguilar A, Cardona L (2014) Paleoindian pinniped exploitation in South America was driven by oceanic productivity. Quat Int 352:85–91. https://doi.org/10.1016/j.quaint.2014.05.015

Schinder G, Guichón RA (2003) Isótopos estables y estilo de vida en muestras óseas humanas de Tierra del Fuego. Magallania 31:33–44

Schinder G, Guichón RA, Comparatore V, Burry S (2003) Inferencias paleodietarias mediante isótopos estables en restos óseos humanos provenientes de Tierra del Fuego, Argentina. Rev Argent Antropol Biol 5(2):15–31

Sealy JC, van der Merwe NJ (1985) Isotope assessment of Holocene human diets in the southwestern Cape, South Africa. Nature 315(6015):138–140. https://doi.org/10.1038/315138a0

Spielmann KA, Eder JF (1994) Hunters and farmers: then and now. Annu Rev Anthropol 23(1):303–323. https://www.jstor.org/stable/2156016

Suby JA (2014a) Nódulos de Schmorl en restos humanos arqueológicos de Patagonia Austral. Magallania 42(1):135–147. https://doi.org/10.4067/S0718-22442014000100008

Suby JA (2014b) Porotic hyperostosis and cribra orbitalia in human remains from Southern Patagonia. Anthropol Sci 122(2):69–79. https://doi.org/10.1537/ase.140430

Suby JA, Guichón RA (2009) Diet, nutrition and femoral robusticity of hunter-gatherers in Southern Patagonia: experiences and perspectives. Int J Osteoarchaeol 19(2):328–336. https://doi.org/10.1002/oa.1040

Suby JA, Zangrando AFJ, Piana EL (2011) Exploraciones osteológicas de la salud de las poblaciones humanas del canal Beagle. Rel Soc Arg Antropol XXXVI:249–270. http://sedici.unlp.edu.ar/handle/10915/20878

Szpak P (2022) Why zooarchaeology needs stable isotope analysis. In: Isotope research in zooarchaeology: methods, applications and advances. University Press of Florida, Boca Raton, pp 248–270

Tafuri MA, Zangrando AFJ, Tessone A, Kochi S, Moggi Cecchi J, Di Vincenzo F, Profico A, Manzi G (2017) Dietary resilience among hunter-gatherers of Tierra del Fuego: isotopic evidence in a diachronic perspective. PLoS One 12(4):e0175594. https://doi.org/10.1371/journal.pone.0175594

Tessone A, Zangrando AFJ, Valencio S, Panarello H (2003) Isótopos estables del carbono en restos óseos humanos de la región del canal Beagle, Isla Grande de Tierra del Fuego. Rev Argent Antropol Biol 5(2):33–43

Tessone A, Guichón RA, Suby JA, Kozameh L (2011) Bioarqueología de Península Mitre. In: Zangrando AFJ, Vázquez M, Tessone A (eds) Los cazadores-recolectores del extremo Oriental fueguino. Arqueología de Península Mitre e Isla de los Estados. Sociedad Argentina de Antropología, Buenos Aires, pp 231–270

Tivoli AM (2012) ¿Intensificación? En el aprovechamiento de aves entre los cazadores- recolectores-pescadores de la región del canal Beagle. Archaeofauna 21:121–137. https://doi.org/10.15366/archaeofauna2012.21.006

Tivoli AM, Zangrando AFJ (2011) Subsistence variations and landscape use among maritime hunter-gatherers. A zooarchaeological analysis from the Beagle Channel (Tierra del Fuego, Argentina). J Archaeol Sci 38(5):1148–1156. https://doi.org/10.1016/j.jas.2010.12.018

Tuhkanen S (1992) The climate of Tierra del Fuego from a vegetation geographical point of view and its ecoclimatic counterparts elsewhere. Acta Bot Fenn 125:4–17

Ulijaszek SJ, Mann N, Elton S (2012) Evolving human nutrition: implications for public health. Cambridge University Press, New York

Valenzuela LO, Laborde PG, D'Angelo del Campo MD, Guichón RA (2019) Transición alimentaria en poblaciones originarias del norte de Tierra del Fuego durante el período misional salesiano: una aproximación isotópica. Rev Argent Antropol Biol 21(1):1–17. https://doi.org/10.17139/raab.2019.0021.01.02

Vales DG, Cardona L, Zangrando AFJ, Borella F, Saporiti F, Goodall RNP, Rosa de Oliveira L, Crespo EA (2017) Holocene changes in the trophic ecology of an apex marine predator in the South Atlantic Ocean. Oecologia 183:555–570. https://doi.org/10.1007/s00442-016-3781-4

Vázquez M, Zangrando AFJ, Tessone A, Ceraso A (2011) Investigaciones arqueológicas recientes en la costa suroriental de Península Mitre. In: Zangrando AF, Vázquez MM, Tessone A (eds) Los cazadores-recolectores del extremo Oriental fueguino. Arqueología de Península Mitre e Isla de los Estados. Sociedad Argentina de Antropología, Buenos Aires, pp 203–230

Vidal HJ (1985) Los conchales de Bahía Valentín. Degree dissertation, Universidad de Buenos Aires

Vidal H (2011 [1985]) El conocimiento de las culturas aborígenes del sudeste de la Isla Grande hasta 1983. In: Zangrando AF, Vázquez MM, Tessone A (eds) Los cazadores-recolectores del extremo Oriental fueguino. Arqueología de Península Mitre e Isla de los Estados. Sociedad Argentina de Antropología, Buenos Aires, pp 65–113

Webb EC, Honch NV, Dunn PJ, Linderholm A, Eriksson G, Lidén K, Evershed RP (2018). Compound-specific amino acid isotopic proxies for distinguishing between terrestrial and aquatic resource consumption. Archaeol Anthropol Sci 10:1–18. https://doi.org/10.1007/s12520-015-0309-5

Whiteman JP, Elliott Smith EA, Besser AC, Newsome SD (2019) A guide to using compound-specific stable isotope analysis to study the fates of molecules in organisms and ecosystems. Diversity 11(1):8. https://doi.org/10.3390/d11010008

Wobst HM (1978) The archaeo-ethnology of hunter-gatherers or the tyranny of the ethnographic record in archaeology. Am Antiq 43(2):303–309. https://doi.org/10.2307/279256

Yesner DR, Figuerero Torres MJ, Guichón R, Borrero LA (1991) Análisis de isótopos estables en esqueletos humanos: confirmación de patrones de subsistencia etnográficos para Tierra del Fuego. Shincal 3:182–191

Yesner DR, Figuerero Torres MJ, Guichón R, Borrero LA (2003) Stable isotopes analysis of human bone and ethnohistoric subsistence patterns in Tierra del Fuego. J Anthropol Archaeol 22:279–291. https://doi.org/10.1016/S0278-4165(03)00040-0

Zangrando AFJ (2009) Historia evolutiva y subsistencia de cazadores-recolectores marítimos de Tierra del Fuego. Sociedad Argentina de Antropología, Buenos Aires

Zangrando AFJ (2018) Poblamiento temprano y arqueología de costas en Patagonia y Tierra del Fuego: vacío de información, preconceptos y perspectivas. Intersecc Antropol 19(2):63–97

Zangrando AFJ, Tessone A, Valencio S, Panarello HO, Salemme M, Mansur ME (2003) Isótopos estables y dietas humanas en ambientes costeros. In: Feliu Ortega MJ, Calleja JM, Edreira Sánchez MC, Fernández Lorenzo MC, Martínez Brell MP, Gil Montero A, Puerto RA (eds) Avances en Arqueometría. Servicio de Publicaciones, Cádiz, pp 91–97

Zangrando AFJ, Tessone A, Vázquez M (2009) El uso de espacios marginales en el archipiélago fueguino: implicaciones de la evidencia arqueológica de Bahía Valentín. In: Salemme M, Santiago F, Álvarez M, Piana E, Vázquez M, Mansur E (eds) Arqueología de la Patagonia: una mirada desde el último confín. Utopías, Ushuaia, pp 47–62

Zangrando AFJ, Panarello HO, Piana EL (2014a) Zooarchaeological and stable isotopic assessments on pinniped–human relations in the Beagle Channel (Tierra del Fuego, Southern South America). Int J Osteoarchaeol 24:231–244. https://doi.org/10.1002/oa.2352

Zangrando AFJ, Tessone A, Ugan A, Gutiérrez MA (2014b) Applications of stable isotope analysis in zooarchaeology: an introduction. Int J Osteoarchaeo 24:127–133. https://doi.org/10.1002/oa.2378

Zangrando AFJ, Riccialdelli L, Kochi S, Nye JW, Tessone A (2016) Stable isotope evidence supports pelagic fishing by hunter-gatherers in southern South America during the Late Holocene. J Archaeol Sci Rep 8:486–491. https://doi.org/10.1016/j.jasrep.2016.05.015

Zangrando AFJ, Kochi S, Nye JW, Tessone A, Riccialdelli L, Tivoli AM, Martinoli MP, Pérez SA, Fogel ML (2022) Long-term interaction between hunter-gatherer-fisher societies and marine resources in the southern tip of South America: applications of stable isotopes in zooarchaeology. In: Sharpe AE, Krigbaum J (eds) Isotope research in zooarchaeology: methods, applications, and advances. University Press of Florida, Gainesville, pp 15–43

Zenteno L, Borella F, Otero JG, Piana E, Belardi JB, Borrero LA, Saporiti F, Cardona L, Crespo E (2015) Shifting niches of marine predators due to human exploitation: the diet of the South American sea lion (*Otaria flavescens*) since the late Holocene as a case study. Paleobiology 41(3):387–401. https://doi.org/10.1017/pab.2015.9

Zhu M, Sealy J (2019) Multi-tissue stable carbon and nitrogen isotope models for dietary reconstruction: evaluation using a Southern African farming population. Am J Phys Anthropol 168(1):145–153. https://doi.org/10.1002/ajpa.23731

Zúñiga Thayer R, Suby J, Luna L, Flensborg G (2020) Osteochondritis dissecans and physical activity in skeletal remains of ancient hunter-gatherers from Southern Patagonia. International J Osteoarchaeol 31(1):77–87. https://doi.org/10.1002/oa.2926

Chapter 6
Expanding Scales in the Human History of the Fuegian Andes

Atilio Francisco J. Zangrando, Daniela V. Alunni, María Paz Martinoli, and Angélica M. Tivoli

Abstract This chapter examines long-term patterns of human mobility, interaction, and adaptation in the Fuegian Andes through an integrated analysis of archaeological, genetic, isotopic, and paleoenvironmental evidence. It explores shifting demographic and cultural dynamics across the Holocene, emphasizing spatial variability in coastal and inland occupation, resource use, and exchange networks among hunter-gatherer-fisher populations. The study highlights significant changes in interaction scales and cultural niche diversity between the Middle and Late Holocene. Evidence suggests that while early populations engaged in expansive networks, later groups experienced more localized interactions, shaped by environmental constraints and social factors. The chapter challenges static ethnographic projections and underscores the complexity of human responses to environmental and demographic shifts, advocating for a nuanced understanding of cultural diversity and resilience in southern Tierra del Fuego.

A. F. J. Zangrando (✉) · D. V. Alunni · M. P. Martinoli · A. M. Tivoli
CADIC-CONICET, Ushuaia, TDFeIAS, Argentina
e-mail: fzangrando@conicet.gov.ar

D. V. Alunni
e-mail: daniela.alunni@conicet.gov.ar

M. P. Martinoli
e-mail: mpmartinoli@conicet.gov.ar

A. M. Tivoli
e-mail: tivoli.angelica@conicet.gov.ar

A. F. J. Zangrando
FFyL-UBA, CABA, Argentina

A. F. J. Zangrando et al. (eds.), *Archaeology of the South Coast of Tierra del Fuego*, The Latin American Studies Book Series, https://doi.org/10.1007/978-3-032-09487-2_6

6.1 Introduction

In previous chapters, we have explored the variability of environmental and demographic conditions along the southern coast of Tierra del Fuego throughout much of the Holocene. We also inferred shifts in interaction networks based on archaeological, genetic, and stable isotope evidence, as well as changes in resource provisioning patterns that may reflect socio-ecological dynamics across time and space. These intertwined processes have shaped a pattern of cultural variability that, despite recent advances, remains only partially understood. In this final chapter, we revisit the key trends identified and critically examine them through the analytical lens of parameters that enable us to explore the diversity of cultural niches.

6.2 The Population Geometric Distribution Changed Over Time

The data analyzed suggest that the geometric distribution of hunter-gatherer-fisher (HGF) populations differed between the Middle and Late Holocene. While both were marked by demographic growth, the nature and extent of social interaction networks varied. During the Middle Holocene, archaeological evidence points to interregional connectivity between the Beagle Channel and those in the southwestern sector of the Patagonian archipelago. In contrast, the Late Holocene is characterized by more geographically restricted interactions in Tierra del Fuego, with increased connectivity between regions south and north of the Fuegian Andes—particularly in the southeastern tip of the island. These patterns suggest that interactions among distinct cultural niches in the region varied over the course of the Holocene.

The demographic increase most robustly documented in the central Beagle Channel during the Middle Holocene appears to be accompanied by a pan-regional distribution of occupations, albeit not necessarily with continuous geographic settlement. This pattern is reflected in the presence of archaeological sites with extensive stratigraphic sequences in various interior areas of the archipelago (e.g., Otway Sound, Strait of Magellan, Beagle Channel), as well as more sporadic evidence in peripheral zones such as Valentín Bay (Península Mitre) (Zangrando et al. 2009) and Ballenero Channel (San Roman et al. 2023) (Fig. 6.1). The distribution of green obsidian across these regions indicates interaction networks connecting distant areas, ranging from locations in the Strait of Magellan and the western Beagle Channel (San Roman et al. 2023) to its eastern mouth at Moat Bay (Zangrando et al. 2021) (Fig. 6.2). Following Borrero et al. (2020), it can be inferred that population density was likely greater at occupation nodes, which also served as hubs of intensified interaction. This, in turn, invites analysis of mobility and circulation strategies among nodes. The presence of green obsidian in remote locations such as the Brecknock Peninsula during the Middle Holocene suggests that these strategies encompassed challenging environments characterized by risky navigation conditions. The limited presence of archaeological sites in the westernmost portion of the Fuegian archipelago suggests a different occupational history compared to other areas, such as the central and eastern Beagle Channel. This pattern implies that the western sector may have been less suitable for long-term settlements, while simultaneously functioning as a key zone for circulation and transit. The extremely rugged terrain of the western Beagle Channel and its proximity to the Darwin Range support the hypothesis that mobility in this region may have involved the use of "porteos" (Borrero et al. 2020). It is also important to consider that during the Middle Holocene, higher sea levels (Rabassa et al. 1990; Gordillo et al. 1992, 1993; Bujalesky 1998, 2007) may have reduced the distances between the head of bays and fjords. These conditions could have facilitated the selection and modification of certain terrestrial sectors for canoe transport, enabling crossings over shorter stretches than those required today.

The conditions that facilitated mobility during the Middle Holocene did not necessarily correspond to densely populated environments. The emergence and widespread distribution of a new lithic projectile point design across the Patagonian archipelago around 4400 cal yr BP appears to illustrate this, as this technology is found in the Beagle Channel despite evidence of low population density. Some scholars interpret this pattern as evidence of the arrival and expansion of maritime hunter-gatherer groups culturally distinct from earlier populations in the region (Morello et al. 2002: 164; Ocampo and Rivas 2004). Others propose that the spread of this type of lithic projectile point reflects the diffusion of a technological concept without requiring population replacement (Piana and Orquera 2007). However, this does not imply that the phenomenon should be interpreted within a single geographical or cultural unit

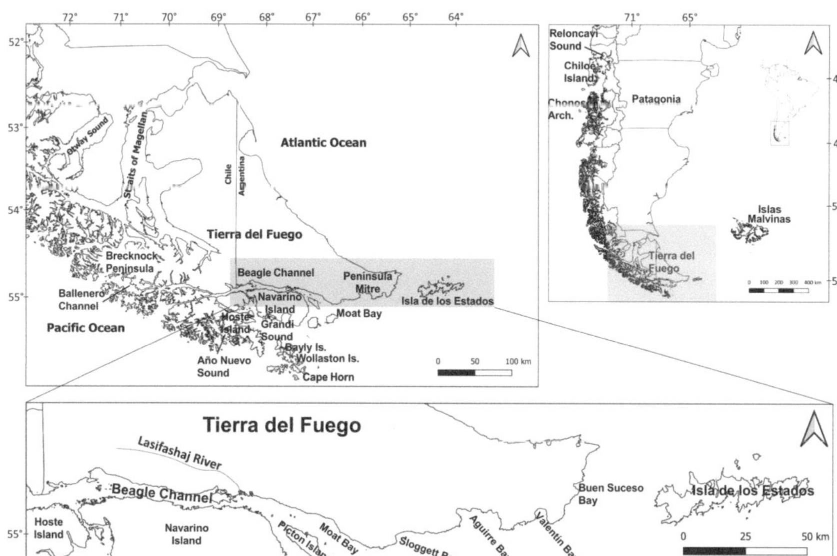

Fig. 6.1 Primary geographical references mentioned in the text

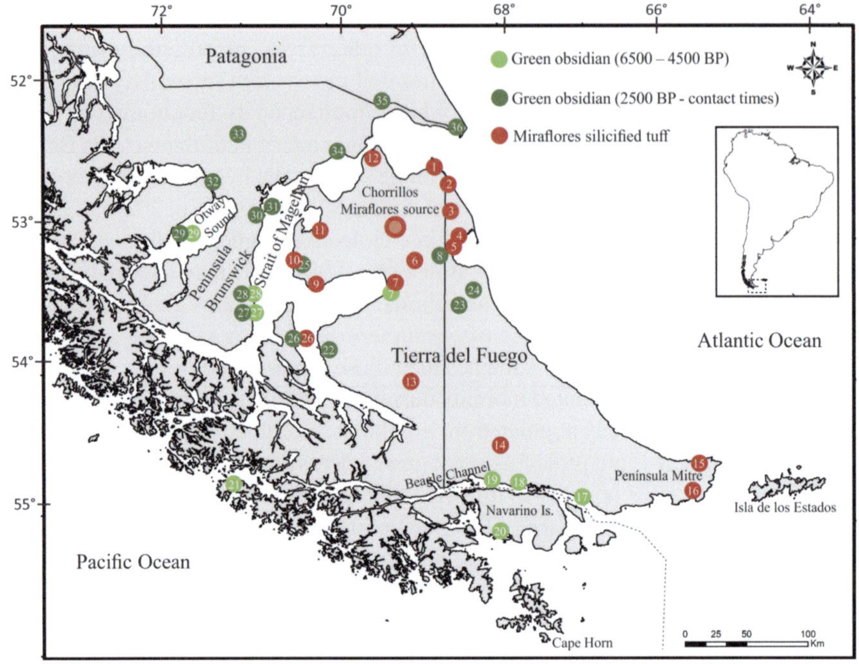

Fig. 6.2 Location of archaeological sites with green obsidian and Miraflores silicified tuff according to Morello et al. (2015), Borrazzo et al. (2019), Zangrando et al. (2021), San Roman et al. (2023): 1. Punta Catalina; 2. Laguna Patria; 3. Río Cullen 2; 4. Laguna al Noroeste de Filaret; 5. Estancias Dos Marías and Florentina; 6. Tres Arroyos; 7. Marazzi; 8. Tres Arroyos 1; 9. Boqueron; 10. Punta Norte 12; 11. Bahía Gente Grande; 12. Punta Baxa; 13. Lago Blanco 12; 14. Kami 1; 15. Caleta Falsa 400; 16. Bahía Valentín 11; 17; Heshkaia 60; 18. Imiwaia I and Binushmuka I; 19. Túnel I; 20. Seno Grandi; 21. Brecknok 1; 22. Río Caleta 4; 23. Amalia 4; 24. Avilés 3; 8. Puesto Consuelo; 25. Porvenir Norte 28; 26. Offing 2; 27.Punta Santa Ana; 28. Punta Carrera; 29. Englefield, Pizzulic 1, 2, 4, Ba. Colorada, Is. Vivian, Englefield sur-tardío; 30. Bahía Laredo 1AC, 3, 5, Cabo Negro, Porpesse, BASHOA 1, 4, 6; 31. Isla Isabel; 32. Río Verde 1; 33. Cañadón Leona 3 y 5; 34. San Gregorio 2; 35. Pali Aike 1 and 2; 36. Dungeness 2, Cabo Vírgenes 6

(Reyes 2021). On the contrary, it is valuable to examine how similar hunting technologies appear in geographically and culturally diverse contexts—such as Reloncaví Seno, the Chonos Archipelago, and the Beagle Channel—including both contexts focused on terrestrial resource exploitation (Orquera et al. 1977; Legoupil 2003) and those oriented toward the hunting of otariids and birds (Legoupil 2023). Comparative technological and functional analyses of these points (Sierralta Navarro 2024) across different contexts may yield important insights into the production and use of this technology in distinct cultural niches.

As discussed in previous chapters, genetic research provides another valuable line of evidence for understanding human interaction in the Fuegian archipelago. The Paiashauaia I mitochondrial haplotype has been assigned to haplogroup D1g (Arencibia et al. 2024), one of the most common haplogroups in Patagonia (Bodner et al. 2012). Bayesian phylogenetic reconstruction indicates that this haplotype is closely related to those of individuals from Navarino Island and Hoste Island, the latter located in Año Nuevo Seno. The estimated age of the most recent common ancestor shared by these three individuals is approximately 4439 years ago (Arencibia et al. 2024). From that point onward, genetic drift may have led to the fixation of different lineages in the Beagle Channel and the islands of the western archipelago (de la Fuente et al. 2018). The genetic affinity between Paiashauaia I and the individual from Hoste Island suggests gene flow or a recent common origin among the groups inhabiting these parts of the archipelago during the Middle Holocene. This finding implies that interaction occurred between these regions during that period. Although the sites where these individuals were recovered are separated by just 90 km, archaeological evidence for the extent of interaction between the Beagle Channel and the southwestern archipelago during the Middle Holocene remains limited. During the historical period, Weddell (2006 [1825]: 159–162) reported the presence of different native groups in the Año Nuevo Seno, while Bridges (1886) excluded the southern sector of Hoste Island from Yagan territory. Several authors have noted that the geographic extent of Yagan populations toward the western end of the archipelago remained uncertain during the nineteenth century—a region that may also have been inhabited by groups ethnically and linguistically related to the Kawésqar (Orquera and Piana 1999a, b). However, as has been proposed for the easternmost part of the Fuegian archipelago (Crespo et al. 2020), the genetic diversity observed in pre-contact populations suggests a higher degree of interaction among groups than is reflected in historical sources.

During the Late Holocene, approximately between 2800 and 500 cal yr BP, there was marked demographic expansion accompanied by a broader geographic distribution of human groups, including the occupation of distant areas such as the Cape Horn archipelago and Isla de los Estados. However, archaeological, paleodietary, and genetic evidence from this period suggests that patterns of the flow of people and information differed from those of earlier phases in the sequence. Evidence suggests that interaction networks become more geographically constrained, primarily within the Tierra del Fuego region, where the configuration of the Fuegian Andes may have influenced and limited the dynamics of population movement and exchange.

As previously discussed in Chap. 2, the Beagle Channel during Late Holocene exhibited notable diversity in lithic projectile point designs. According to some scholars, these forms show parallels with those found in archaeological assemblages along the Atlantic coast (Piana and Orquera 2007). Others, however, argue that morphometric similarities with northern Tierra del Fuego are largely confined to Península Mitre, with no clear correspondences between the Beagle Channel and other regions of the island (Charlin et al. 2014). Despite these differing interpretations, the evidence supports the notion of information flow between the southern coast of Tierra del Fuego and more northern parts of the island. This interpretation is further supported by the distribution of non-local raw materials. While green obsidian reappears with some frequency in Late Holocene assemblages from the Strait of Magellan region, it is largely absent from those in the Beagle Channel (Morello et al. 2012).

Nevertheless, evidence of silicified tuff transport to sites in Península Mitre suggests connections with northern Tierra del Fuego (Pallo and Borrazzo 2016) (Fig. 6.2).

Kochi (2023) examined the mobility of 69 individuals based on oxygen isotopic values ($\delta^{18}O_c$) from the carbonate of skeletal remains found on the southern coast of Tierra del Fuego and Isla de los Estados. The study spans a chronological range from approximately 2500 cal yr BP to the period of European contact. This stable isotope proxy enabled the identification of atypical cases of individuals exhibiting foraging ranges distinct from the rest of the sample, or potentially foreign individuals. The modern $\delta^{18}O$ reference frame for water sources reveals a longitudinal gradient, with the most enriched waters found to the east of Isla Grande, while no latitudinal differences are observed in relation to the northern regions of Tierra del Fuego. The application of this method suggests that the proportion of non-local individuals is very low, which aligns with the idea of limited interaction with more northern or western sectors of the archipelago. At the same time, it may reflect the persistence of social networks maintained through visitation mechanisms between different areas of Tierra del Fuego (Kochi 2023). This idea is similar to that of Pallo and Borrazzo (2016), who suggested the existence of north–south circulation routes or "visits" to the southern coast of Península Mitre, based on the irregular distribution of silicified tuff. Kochi (2023) similarly argues that the $\delta^{18}O_c$ isotopic data do not exclude the possibility of visits or other social interaction mechanisms originating from the north or other regions.

Returning to the discussion in Chap. 5, Kochi (2023) highlights that the primary pattern revealed by oxygen isotope analysis is the integration of similar mobility ranges throughout the lives of individuals from the southern coast of Isla Grande de Tierra del Fuego, Península Mitre, and Isla de los Estados. Despite the presence of a positive $\delta^{18}O$ gradient in water sources toward the east of Isla Grande, $\delta^{18}O_c$ values in human remains do not show clear geographic segregation. Moreover, the estimated drinking water values suggest that the majority of human isotopic signatures can be explained by a combination of sources from the central and eastern portions of the Beagle Channel, as well as the more enriched waters at the island's eastern end. This isotopic evidence thus supports the hypothesis of shared population dynamics encompassing the Beagle Channel, Península Mitre, and Isla de los Estados (Zangrando et al. 2009). As discussed in Chap. 2, genetic studies further reinforce the existence of interaction among these areas, with shared maternal lineages between groups inhabiting the southern coast of Isla Grande and Isla de los Estados (Crespo et al. 2020). However, the genetic evidence also complicates the picture of occupation in Península Mitre, as some individuals share haplotypes not only with ancient Yagan populations but also with groups from more northerly regions. Based on this evidence, Crespo et al. (2020) propose the occurrence of recent gene flow events or a recent common origin with populations from northern Tierra del Fuego or continental Patagonia.

In summary, these various lines of evidence suggest that interaction with areas north of the Fuegian Andes was more pronounced at the southeastern tip of Tierra del Fuego. This supports the hypothesis that Península Mitre functioned as a permeable corridor for the movement of human groups across a broader landscape (Lanata 2002;

Kochi 2023; Vázquez 2024)—a dynamic less evident in the Beagle Channel. These findings prompt a reconsideration of cultural diversity during the Late Holocene, moving beyond simplistic dichotomies of marine versus terrestrial subsistence to include groups exhibiting "mixed" mobility patterns, as recorded in historical accounts (Zangrando et al. 2014a, b). A key dimension of this discussion involves analyzing the dynamics between coastal and interior areas.

6.3 Coast-Inland Interaction Varied Spatially Along the Southern Coast of Tierra del Fuego

To analyze the interaction between coastal and inland environments along the southern coast of Tierra del Fuego, a key initial consideration is that most known occupations are primarily associated with the coast. This pattern is evident, for instance, along the northern coast of the Beagle Channel: after identifying a high density of coastal sites, archaeological evidence becomes virtually absent inland, with only a few exceptions related to the procurement of lithic raw materials (e.g., Álvarez et al. 2010). A similar situation is observed in Península Mitre, where the abundant archaeological record along its coastal perimeter (Zangrando et al. 2011; Vázquez 2024) contrasts with the current absence of documented inland sites. Consequently, analyses of coast-interior interactions must currently rely on indirect evidence derived from indicators present in the coastal archaeological record. A second point to consider concerns the different dimensions of mobility involved in coast-interior interaction. Following the framework proposed by MacDonald and Hewlett (1999), a narrower spatial dimension of mobility is primarily associated with the procurement of resources necessary for daily subsistence. In contrast, a broader dimension encompasses movements over intermediate distances aimed at establishing interpersonal relationships—such as those related to marriage exchanges. Beyond these, there is an even larger-scale dimension of mobility, which may involve long-distance travel for the acquisition of exotic goods, a type of movement addressed in the previous discussion.

To evaluate the first dimension of mobility—comparable to concepts such as *site catchment* (Vita-Finzi et al. 1970) or *foraging radius* (Binford 1983)—it is essential to consider the constraints imposed by the topography of the Fuegian Andes. The range decreases in elevation from west to east, producing increasingly broad and accessible valleys perpendicular to the coastline as one approaches Península Mitre (Fig. 6.3). This geographical configuration influenced not only inland human mobility but also the availability and accessibility of terrestrial resources, such as guanaco (Alunni and Zangrando 2012). As outlined in Chap. 3, while the spatial distribution of guanaco remains in archaeological assemblages does not vary significantly over time during the Holocene, there is clear spatial variability in the strategies used for transporting and processing this resource (Alunni 2021), including the exploitation of by-products such as bone fat (Alunni et al. 2021). Notably, evidence of labor-intensive bone fat extraction is found along the coast of the Beagle Channel but is absent in Península Mitre, suggesting differences in subsistence strategies and possibly in local ecological or logistical conditions. This pattern may be explained not only by the geographic characteristics of Península Mitre—which likely facilitated easier access to interior zones—but also by the potential year-round coastal presence of guanaco herds. For instance, guanaco groups have been repeatedly observed in Valentín Bay during the summer months (Vázquez 2024). Additionally, the presence of numerous otariids colonies along the Península Mitre coastline offered alternative fat-rich resources. Based on trends derived from the analysis of multiple archaeological assemblages, it can be inferred that guanaco exploitation in the Beagle Channel, where the resource was less readily available, may have required broader and more costly provisioning ranges compared to Península Mitre. This could partially explain the more intensive processing of guanaco remains observed in the former region. Ethnographic data support this interpretation by the late nineteenth century: Yagan groups reportedly hunted guanacos primarily in winter, organizing multi-day expeditions involving several individuals who traveled into inland valleys in search of the animals (Orquera and Piana 1999b: 140–142). However, this information may not accurately represent conditions during pre-contact times. Environmental stress linked to the severe climatic conditions of the Little Ice Age could have significantly impacted guanaco availability in both inland and coastal zones along the Beagle Channel (Fernández et al. 2021). This interpretation is further supported by recent genetic studies (Leggieri et al. 2024). We will revisit the potential effects of this climatic event on human populations along the southern coast of Tierra del Fuego in a subsequent section.

Kochi (2023), in evaluating protein intake through $\delta^{13}C_{col}$ and $\delta^{15}N$ values in individuals from the Beagle Channel and Península Mitre, observed overlapping central trends across different age and sex categories in both regions. However, in Península Mitre, adult males tend to exhibit more negative values for both isotopic markers. The author tentatively suggests that these patterns may reflect a greater consumption of terrestrial prey by several adult males in Península Mitre, indicating foraging ranges with stronger connections to terrestrial environments. In contrast, the similar and tightly clustered $\delta^{13}C_{col}$ and $\delta^{15}N$ values observed among adult females suggest long-term consumption of more homogeneous, primarily marine diets (Kochi 2023). These findings imply that inland provisioning strategies may also be traceable at the individual level in Península Mitre.

Another notable finding, which relates to broader dimensions of mobility, is that paleodietary diversity is greater along the northern coast of Península Mitre than along its southern coast (Kochi 2023). Moreover, the isotopic niche of the Beagle Channel nearly completely overlaps with that of the southern coast of the peninsula and Isla de los Estados, suggesting long-term similarities in subsistence strategies across this corridor. In contrast, the northern coast of Península Mitre exhibits greater paleodietary variability, likely reflecting more diverse life histories. The proximity of this northern isotopic niche to that of groups from northern Isla Grande de Tierra del Fuego further implies closer connections with those populations (Kochi 2023). Vázquez (2024) reached a similar conclusion based on historical and ethnographic

Fig. 6.3 View of the Fuegian Andes from the eastern headwaters of Lake Kami (or Fagnano). At the center of the image, a broad mountain pass known as Senda Bridges—formerly used by Indigenous peoples—is prominently visible

sources. Together, the stable isotope data and detailed analysis of written records challenge the projection of colonial-era ethnic divisions onto pre-contact contexts. Instead, these multiple lines of evidence point to a mosaic of distinct cultural niches.

6.4 Isolation and Multi-island Interaction

As highlighted in Chap. 1, the concept of isolation should not be conflated with enclosure or confinement. Recent studies have demonstrated that islands are not closed systems; on the contrary, analyzing the nature, extent, timing, and duration of isolation is fundamental to understanding patterns of island occupation (Fitzhugh 1997; Memmott et al. 2006). This perspective emphasizes the importance of analyzing islands in relation to continental areas or other islands, as their occupational histories only gain full meaning within such broader spatial contexts (Boomert and Bright 2007). Referring to isolation as a relative concept (Broodbank 2008; Erlandson 2008) involves not only considering enabling factors such as the presence of navigational technology for island exploration and colonization (Fitzpatrick and Atholl 2008), but also recognizing that certain environments within an archipelago impose greater

Fig. 6.4 View of Isla de los Estados from the eastern tip of the Península Mitre

constraints and uncertainties on human occupation than others. In this context, the geographic position of Isla de los Estados—due to its offshore location—differs significantly from that of the Beagle Channel and the Península Mitre. Navigating the waters around Isla de los Estados remains one of the most formidable nautical challenges in the region today (Fig. 6.4).

In previous chapters, we analyzed the occupation patterns of this island based on chronological data and the composition of archaeological deposits. The results suggest that, regardless of whether particular areas were used regularly as part of a seasonal cycle—either for residential or logistical purposes—the archaeological record of Isla de los Estados presents features that differ markedly from those documented in other outer zones of the Tierra del Fuego archipelago, as evidenced by research at Cape Horn (Legoupil 1993–94). Despite notable contrasts with the high density of occupations in Grandi Seno (southern Navarino Island), the occupational redundancy documented at numerous shell midden sites on Bayly and Wollaston Islands suggests that "*a very small but stable population would have gradually established itself in the Cape Horn archipelago following undoubtedly more seasonal incursions, dating back at least to the beginning of our era*" (Legoupil 1993–94: 119, our translation). On the other hand, the occupations recorded in Isla de los Estados seem to reflect another form of use that could be understood as part of extended ranges (Binford 1982: 8) of population nodes located in other sectors of the archipelago perhaps as cases of exploration without detachment from an original population (Borrero 2001: 130; Zangrando et al. 2024). In addition to this situation, no archaeological occupations have been detected on the island between 1375 cal yr BP and the arrival of Europeans to the region, which has been interpreted as a period of abandonment (Horwitz 1990). This pattern contrasts notably with the evidence from Cape Horn, which maintained population continuity during the final stage of the Holocene.

This shows that the cultural niches of HGF populations did not integrate offshore spaces with the same intensity throughout the Holocene. Such variability may reflect different forms of social interaction, whether due to geographic distance from population nodes, demographic factors, presence of other populations, among other situations (Irwin 1992; Weisler 1995; Anderson 2004).

6.5 Resource Fluctuations and Resilience

As noted in Chap. 1, resilience analysis provides insights into the ability of human populations to absorb or adapt to stressful conditions in their social or natural environments. The analysis presented in Chap. 3 demonstrates how subsistence patterns varied over time, based on the representation of faunal resources throughout the archaeological sequence. One of the most notable changes involves the representation of pinnipeds in the zooarchaeological assemblages. A general trend shows a decline in pinniped remains around 5300 cal yr BP, coinciding with an increase in alternative resources such as guanacos, birds, and marine fish. This decline is also associated with shifts in the size composition of exploited prey, indicated by a greater proportion of females and juveniles, suggesting changes in encounter rates and provisioning strategies (Zangrando et al. 2014a, b; Martinoli 2018). As discussed in Chap. 5, stable isotope analyses ($\delta^{13}C$ and $\delta^{15}N$) of pinniped bone collagen from different periods in the sequence further suggest changes in pinniped isotopic niches over time. Although no significant variations were observed in the trophic levels of *Arctocephalus australis* and *Otaria flavescens*, both species exhibited shifts in their foraging areas—from coastal waters to more offshore zones—between the Middle and Late Holocene (Zangrando et al. 2014a, b; Nye et al. 2020). While these changes do not appear to correlate directly with climatic fluctuations during the periods analyzed, they do coincide with a decline in pinniped representation in the archaeological record, a reduction in the size of exploited individuals, and an increase in site occupation frequency around 5300 cal yr BP. Although the co-variation of these trends does not imply a direct causal relationship, it is plausible to hypothesize that intensified human predation pressured pinnipeds to forage and inhabit areas farther from the coast (Nye et al. 2020). As the cost of capturing these animals increased, hunter-gatherer-fisher populations responded by diversifying their subsistence strategies. This process has been interpreted as a form of niche broadening, which may have enhanced the regional carrying capacity for these populations (Tivoli and Zangrando 2011).

Resource fluctuations can lead to periods of famine (Borrero et al. 2020). The ability to cope with periods of resource scarcity is related to the resilience of populations (Tafuri et al. 2017). Historical accounts confirm that the Yagan experienced episodes of famine during the nineteenth century, particularly in the context of European contact and colonial expansion (Orquera and Piana 1999b: 185–190). These situations were mainly recorded in the eighteenth and nineteenth centuries and may have been due to a confluence of social and environmental factors. In the first place, the overexploitation of pinnipeds in the region by North American and European factories to obtain fur and blubber throughout those centuries may have had an impact on the possibilities of obtaining this resource by HGF populations (Orquera 2002). In addition to this situation, there was a significant climatic cooling event known as the Little Ice Age that lasted between the fourteenth and nineteenth centuries in the region. This climatic period had a profound impact on guanaco populations in Tierra del Fuego. This pattern is evidenced not only by a marked decline in the representation of this resource in the archaeological record (Zangrando 2009; Fernández et al. 2021), but also by recent genetic studies indicating that guanaco populations underwent bottlenecks during this period (Leggieri et al. 2024). The confluence of both processes suggests that both resources (pinnipeds and guanacos) became less available, thereby limiting access to fat—a critical nutrient for human subsistence in this region. This extended period of resource scarcity was likely interrupted only occasionally by whale strandings (Borrero et al. 2020), events whose importance to indigenous populations is well-documented in both historical and ethnographic sources (Orquera and Piana 1999b). Such circumstances may have fostered the development of social strategies involving aggregation and exchange among groups, as well as the deferred consumption of meat and blubber through storage practices (Orquera and Piana 1999b: 138) (Fig. 6.5). While these behaviors have been proposed for the nineteenth century based on historical and archaeological evidence (Piana et al. 2000; Briz et al. 2009), they have yet to be confirmed for earlier periods of the Holocene.

High human population density in the Beagle Channel region is primarily inferred from the abundance of archaeological sites documented in the area (Orquera and Piana 1999a; Barceló et al. 2002; Eidshaug et al. 2024). However, as discussed in Chap. 2, this density should not be considered constant throughout the second half of the Holocene. Our analysis of the Beagle Channel indicates that, following a significant population increase during the Late Holocene—as inferred from summed probability distributions of radiocarbon dates—a marked decline in site occupation is observed on the northern coast of the Beagle Channel over the past 500 years. In contrast, demographic trends in Peninsula Mitre appear stable over the same period.

Observations made during the early nineteenth century voyages of the HMS Adventure and HMS Beagle (1826–1836), which surveyed both the coasts of Peninsula Mitre and the central Beagle Channel, are consistent with our paleodemographic models. While encounters with Indigenous groups—whether during landings or via sightings from the ship—were frequent along many coastal areas of Peninsula Mitre and the eastern islands (Parker King 2015 [1839]: 383–884; Fitz Roy 2016 [1839]: 122–123; Darwin 2019 [1839]: 214), such encounters were notably absent or less

Fig. 6.5 Wood engraving showing the use of a beached whale in Tierra del Fuego (1860, Klaus Barthelmess Collection, Cologne, Germany, No. 1234; In Mayne Reid, T., 1860, p. 424; engraving by J. B. Zwecker and Dalziel Brothers. This image was reproduced by several authors (see Quiroz et al. 2016)

frequent in the central Beagle Channel. For example, in January 1833, Fitz-Roy sailed along the northern coast of the Beagle Channel between Gable Island and Murray Channel without recording any encounters with Indigenous groups (Fitz-Roy 2016 [1839]: 186). As reported by Yagan informant Orundellico (Jemmy Button), who was aboard the HMS Beagle, this area was commonly frequented by terrestrial Indigenous groups from north of the Fuegian Andes. Orundellico's testimony suggests that these groups conducted expeditions—sometimes extending south of the Beagle Channel—for the purpose of looting (Fitz-Roy 2016 [1839]: 187–188). The conditions observed during the seventeenth and nineteenth centuries, particularly the scarcity of pinnipeds and guanacos, may have compelled indigenous groups to seek alternative resources, prompting shifts in patterns of mobility and spatial occupation. Additionally, the arrival of early European ships—initially confined to areas outside the interior channels—may have influenced Indigenous dynamics (Borrero 2001; Vázquez 2024). The impact of European presence must be taken into account when interpreting the historical trajectories of native populations in the region. The establishment of the Anglican Mission and other European settlements toward the end of the nineteenth century offers clear evidence of these effects, including the aggregation of several hundred individuals in specific locations (Orquera and Piana 1999b: 187).

6.6 Cultural Niche Diversity as a Prospect

The cultural niche model applied in this book enabled the integration of environmental, archaeological, genetic, and stable isotope data within a broad spatio-temporal framework. This approach reveals that the history of HGF populations along the southern coast of Tierra del Fuego was shaped by both natural and social dynamics. These include fluctuations in resource availability, demographic shifts, and changes in mobility and interaction patterns. The results suggest the existence of multiple cultural niches throughout the Holocene, rather than a single, stable mode of adaptation to the coastal environment across the regional archaeological sequence.

The documented variability also highlights the limitations of relying on a stereotypical interpretation derived from the ethnographic record to understand the spatial organization of hunter-gatherer-fisher populations in the Fuegian region. Not only is it problematic to project this model into the deep past—given that archaeological and genetic evidence reflects long-term population dynamism throughout the Holocene—but ethnographic and historical data themselves suggest that the configuration of these populations' cultural niches could change significantly over the span of just a few centuries.

This perspective opens new directions for regional archaeological research. For instance, the data presented here do not necessarily support continuous population presence in the Beagle Channel over the past 8600 cal yr BP. While it is important to consider potential sampling biases—particularly given that research has often focused on shell midden excavations—the periods with fewer or no occupations, which may reflect episodes of population collapse, could signal the limits of adaptive plasticity among HGF groups. Similar patterns may also be observed on smaller spatial scales, such as the interruption of occupations on Isla de los Estados. Discontinuities are also evident in spatial patterns. There is no reason to dismiss the possibility that social interactions between the coastal regions of Tierra del Fuego and the interior were particularly active during the Late Holocene, as suggested by ethnographic sources. However, our understanding of these dynamics remains limited (Borrero et al. 2020). For instance, no archaeological evidence has been recorded in broad valleys such as the Lasifashaj River valley, which extends a few kilometers inland while running parallel to the Beagle Channel, suggesting that it could have been an important route for circulation. While this absence may partly result from the lack of systematic research in these areas, it could also be explained by other factors that account for the presence of unoccupied spaces within the movement dynamics of HGF populations (Borrero et al. 2020). It is possible that this mountain range sector acted as a social buffer between groups, thereby reducing potential conflict (Eerkens 1999).

A strong influence persists in the region from the use of categories such as "littoral," "coastal," and "terrestrial," which oversimplify the long-term history of HGF populations and marginalize the cultural dynamics between groups with distinct ways of life (Zangrando et al. 2014a, b). A deeper understanding of how, when, and why these processes unfolded will enable us to explain how HGF shaped their social environments to control, influence, and even reverse the flow of people, ideas, and material goods across different scales. In comparison to other regions of the world, we argue that Tierra del Fuego presents a unique opportunity for conducting significant research into these processes.

References

Alunni DV, Zangrando AF (2012) Primeros datos sobre el transporte, procesamiento y consumo de guanacos en la localidad arqueológica Heshakaia (sudeste de Tierra del Fuego, Argentina). Magallania 40(1):319–331. https://doi.org/10.4067/S0718-22442012000100019

Alunni DV (2021) Ungulates butchering and transport by hunter gatherers with maritime economic orientation: the case of the south coast of Tierra del Fuego (Argentina). J Anthropol Archaeol 64:101336. https://doi.org/10.1016/j.jaa.2021.101336

Alunni DV, Kochi S, Tessone A, Zangrando AF (2021) Consumo de grasas entre cazadores-recolectores de la costa sur de Tierra del Fuego durante el Holoceno tardío: una evaluación desde la zooarqueología y los isótopos estables. Comechingonia 25(3):61–70

Álvarez M, Pal N, Salvatelli L, Briz Godino I, Zangrando AF, Bazerque A, De Angelis H, Fiore D (2010) Tallando desde lo alto: un sitio de explotación de materiales líticos de los grupos cazadores-recolectores de la Isla Grande de Tierra del Fuego. Magallania 38(1):297–302. https://doi.org/10.4067/S0718-22442010000100019

Anderson AJ (2004) Islands of ambivalence. In: Fitzpatrick SM (ed) Voyages of discovery: the archaeology of islands. Praeger, Westport CT, pp 251–273

Arencibia V, Muñoz Hidalgo M, Crespo M, Maldonado L, Lichtenstein G, Kamenetzky L, Vera P, Zangrando AF, Tessone A, Avena S, Lia V, Puebla A, Dejean C (2024) Ancient human mitogenome of the Beagle Channel (Tierra del Fuego): an Argentine collaborative project. Lat Am Antiq 35(1):149–162. https://doi.org/10.1017/laq.2023.13

Barceló JA, Piana EL, Martinioni DR (2002) Archaeological spatial modelling: a case study from Beagle Channel (Argentina). In: Burenhult G, Arvidsson J (eds) Archaeological informatics: pushing the envelope, computer applications and quantitative methods in archaeology. BAR series 1016, Oxford, pp 351–360

Binford LR (1982) The archaeology of place. J Anthropol Archaeol 1:5–31. https://doi.org/10.1016/0278-4165(82)90006-X

Binford LR (1983) Long term land use patterns: some implications for archaeology. In: Dunnell RC, Grayson DK (eds) Lulu linear punctated: essays in honor of George Irving Quimby. University of Michigan, Ann Arbor, Anthropological papers N° 72, pp 27–53

Bodner M, Perego UA, Huber G, Fendt L, Röck AW, Zimmermann B, Olivieri A, Gómez-Carballa A, Lancioni H, Angerhofer N, Bobillo MC, Corach D, Woodward SR, Salas A, Achilli A, Torroni A, Bandelt HJ, Parson W (2012) Rapid coastal spread of First Americans: novel insights from South America's Southern Cone mitochondrial genomes. Genome Res 22(5):811–820. https://doi.org/10.1101/gr.131722.111

Boomert A, Bright AJ (2007) Island archaeology: in search of a new horizon. Island Stud J 2(1):3–26. https://doi.org/10.24043/isj.197

Borrazzo KB, Borrero LA, Pallo MC (2019) Exploring lithic transport in Tierra del Fuego (Southern South America). J Archaeol Sci Rep 24:220–230. https://doi.org/10.1016/j.jasrep.2019.01.002

Borrero LA (2001) El poblamiento de la Patagonia. Toldos, Milodones y Volcanes. Editorial Emecé, Buenos Aires

Borrero LA, Morello F, San Román M (2020) Circulación de bienes, uso del espacio interior y espacios programados en los archipiélagos de Fuego-Patagonia en tiempos recientes. Magallania 48(2):71–98. https://doi.org/10.4067/S0718-22442020000200071

Bridges T (1886) El confín de la República: la Tierra del Fuego y sus habitantes. Bol Ins Geogr Argent VII:313–320
Briz I, Álvarez M, Zurro D, Caro J (2009) Meet for lunch in Tierra del Fuego: a new ethnoarchaeological project. Antiquity 83:322
Broodbank C (2008) Not waving but drowning. J Island Coast Archaeol 3:72–76. https://doi.org/10.1080/15564890801990771
Bujalesky G (1998) Holocene coastal evolution of Tierra del Fuego, Argentina. Quat S Am Antarct Penins 11:247–281
Bujalesky G (2007) Coastal geomorphology and evolution of Tierra del Fuego (Southern Argentina). Geol Acta 5(4):337–362
Crespo CM, Cardozo DG, Tessone A, Vázquez MM, Kisielinski C, Arencibia V, Tackney J, Zangrando AF, Dejean CB (2020) Distribution of maternal lineages in hunter-gatherer societies of the southern coast of Tierra del Fuego, Argentina. Am J Phys Anthropol 173(4):709–720. https://doi.org/10.1002/ajpa.24107
Charlin J, Cardillo M, Borrazzo K (2014) Spatial patterns in Late Holocene lithic projectile point technology of Tierra del Fuego (Southern South America): assessing size and shape changes. World Archaeol 46(1):78–100. https://doi.org/10.1080/00438243.2014.890914
Darwin C (2019 [1839]) Los viajes del Beagle. Diario y observaciones (1832–1836). Colección Reservada del Museo del Fin del Mundo. Eudeba, Buenos Aires
de la Fuente C, Ávila-Arcos MC, Galimany J, Carpenter ML, Homburger JR, Blanco A, Contreras P, Dávalos DC, Reyes O, Roman MS, Moreno-Estrada A, Campos PF, Eng C, Huntsman S, Burchard EG, Malaspinas AS, Bustamante CD, Willerslev E, Llop E, Verdugo RA, Moraga M (2018) Genomic insights into the origin and diversification of late maritime hunter-gatherers from the Chilean Patagonia. Proc Natl Acad Sci USA 115(17):E4006–E4012. https://doi.org/10.1073/pnas.1715688115
Eerkens JW (1999) Common pool resources, buffer zones, and jointly owned territories: hunter-gatherer land and resource tenure in Fort Irwin, Southeastern California. Hum Ecol 27:297–318. https://doi.org/10.1023/A:1018777311943
Eidshaug JSP, Bjerck HB, Breivik HM, Risbøl O, Tivoli AM, Zangrando AF (2024) From earth to sky: large-scale archaeological settlement patterns in southernmost South America based on ground surveys, UAV LiDAR, and open access satellite imagery. J Island Coast Archaeol 1–40. https://doi.org/10.1080/15564894.2024.2426828
Erlandson J (2008) Isolation, interaction, and island archaeology. J Island Coast Archaeol 3:83–86. https://doi.org/10.1080/15564890801999939
Fernández M, Ponce FJ, Zangrando AF, Borromei AM, Musotto LL, Alunni DV, Vázquez M (2021) Relationships between terrestrial animal exploitation, marine hunter-gatherers and palaeoenvironmental conditions during the Middle-Late Holocene in the Beagle Channel region (Tierra del Fuego). Quat Int 549:208–217. https://doi.org/10.1016/j.quaint.2018.05.032
Fitzhugh W (1997) Biogeographical archaeology in the Eastern North American Arctic. Hum Ecol 25(3):385–418. https://doi.org/10.1023/A:1021819509181
Fitzpatrick SM, Atholl A (2008) Islands of isolation: archaeology and the power of aquatic perimeters. J Island Coast Archaeol 3(1):4–16. https://doi.org/10.1080/15564890801983941
Fitz Roy R (2016 [1839]) Los viajes del Beagle. Informes de la Segunda Expedición (1831–1836). Colección Reservada del Museo del Fin del Mundo. Eudeba, Buenos Aires
Gordillo S, Bujalesky G, Pirazzoli PA, Rabassa JO, Saliège JF (1992) Holocene raised beaches along the northern coast of the Beagle Channel, Tierra del Fuego, Argentina. Palaeogeogr Palaeoclimatol Palaeoecol 99:41–54. https://doi.org/10.1016/0031-0182(92)90006-Q
Gordillo S, Coronato AMJ, Rabassa JO (1993) Late Quaternary evolution of a subantarctic paleofjord, Tierra del Fuego. Quat Sci Rev 12:889–897. https://doi.org/10.1016/0277-3791(93)90027-J
Horwitz V (1990) Maritime settlement patterns in Southeastern Tierra del Fuego (Argentina). Ph.D. dissertation, University of Kentucky

Irwin GJ (1992) The prehistoric exploration and colonization of the Pacific. Cambridge University Press, Cambridge

Kochi S (2023) El nicho ecológico de los cazadores-recolectores del canal Beagle, Península Mitre e Isla de los Estados durante el Holoceno tardío y sus implicancias socioculturales. Ph.D. dissertation, Universidad de Buenos Aires

Lanata JL (2002) The world's southernmost foragers: the native diversity of Tierra del Fuego. In: Briones C, Lanata JL (eds) Archaeological and anthropological perspectives on the native peoples of Pampa, Patagonia, and Tierra del Fuego to the nineteenth century. Bergin & Garvey, London, pp 57–73

Leggieri LR, Anello M, Peralta D, Túnez JI, Di Rocco F, Poljak S, Flores C, Alunni D, Belardi J B, Gutiérrez MA, Álvarez MC, Massigoge A, Kaufmann C, Borrero L, Borghi C, Demartini J, Petracci P, Martín S, Marbán L, Kathiravan P, Rudolf P, Marín JC, Cárcamo JG, Caramachai P (2024) Delineating genomic features for wild guanaco conservation. Biol J Linn Soc 143(3):blae087. https://doi.org/10.1093/biolinnean/blae087

Legoupil D (1993–94) El archipiélago del cabo de Hornos y la costa sur de la isla Navarino: poblamiento y modelos económicos. Anal Inst Patag Ser Cs Hum 22:101–121

Legoupil D (ed) (2003) Cazadores-recolectores de Ponsonby (Patagonia austral) y su paleoambiente desde VI al III milenio AC. Universidad de Magallanes, Punta Arenas

Legoupil D (2023) Entre Patagonia y Tierra del Fuego: Los nómadas del mar del islote Offing (I Colección Poblamiento Humano de Fuego-Patagonia. Ediciones Universidad de Magallanes, Punta Arenas Dawson - Estrecho de Magallanes)

MacDonald DH, Hewlett BS (1999) Reproductive interests and forager mobility. Curr Anthropol 40:501–523. https://doi.org/10.1086/200047

Martinoli MP (2018) Modalidades de explotación, procesamiento y consumo de pinnípedos en la margen meridional de Tierra del Fuego. Ph.D. dissertation, Universidad de Buenos Aires

Memmott P, Evans N, Robins R, Lilley I (2006) Understanding isolation and change in island human populations through a study of indigenous cultural patterns in the Gulf of Carpentaria. Trans R Soc S Aust 130(1):29–47. https://doi.org/10.1080/3721426.2006.10887046

Morello F, Borrero LA, Massone M, Stern C, Garcia-Herbst A, McCulloch R, Arroyo-Kalin M, Calás E, Torres J, Prieto A, Martinez I, Bahamonde G, Cárdenas P (2012) Hunter-gatherers, biogeographic barriers and the development of human settlement in Tierra del Fuego. Antiquity 86:71–87. https://doi.org/10.1017/S0003598X00062463

Morello F, San Román M, Prieto A (2002) Puntas de proyectil lanceoladas en Patagonia Meridional y Tierra del Fuego. Anal Inst Patag Ser Cs Hum 30:155–166

Morello F, Stern CH, San Román M (2015) Obsidiana verde en Tierra del Fuego y Patagonia: caracterización, distribución y problemáticas culturales a lo largo del Holoceno. Intersecc Antropol 2:139–153

Nye JW, Zangrando AF, Martinoli MP, Fogel ML (2020) Temporal and population trends in human exploited pinnipeds from Tierra del Fuego. Palaeogeogr Palaeoclimatol Palaeoecol 55:4109804. https://doi.org/10.1016/j.palaeo.2020.109804

Ocampo C, Rivas P (2004) Poblamiento temprano de los extremos geográficos de los canales patagónicos: Chiloé e Isla Navarino 1. Chungara 36:317–331. https://doi.org/10.4067/S0717-73562004000300034

Orquera LA (2002) The late nineteenth-century crisis in the survival of the Magellan-Fueguian Litoral Natives. In: Briones C, Lanata JL (eds) Archaeological and anthropological perspectives on the native peoples of Pampa, Patagonia, and Tierra del Fuego to the nineteenth century. Bergin & Garvey, London, pp 145–158

Orquera LA, Piana EL (1999a) Arqueología de la región del canal Beagle (Tierra del Fuego, República Argentina). Sociedad Argentina de Antropología, Buenos Aires

Orquera LA, Piana EL (1999b) La vida material y social de los Yámana. Eudeba, Buenos Aires

Orquera LA, Sala A, Piana EL, Tapia A (1977) Lancha Packewaia. Arqueología de los Canales Fueguinos. Temas de Arqueología. Huemul, Buenos Aires

Pallo MC, Borrazzo K (2016) Miraflores rocks in Tierra del Fuego (southernmost South America): hunter-gatherer procurement, optimal pathway analysis and social interaction. J Archaeol Sci Rep 7:420–431. https://doi.org/10.1016/j.jasrep.2016.05.024

Parker King P (2015 [1839]) Los viajes del Beagle. Informe de la Primera Expedición (1826–1830). Colección Reservada del Museo del Fin del Mundo. Eudeba, Buenos Aires

Piana EL, Estévez Escalera J, Vila Mitjá A (2000) Lanashuaia: un sitio de canoeros del siglo pasado en la costa norte del canal Beagle. In: Carballo Marina F, Espinosa S, Belardi JB, De Nigris M, Fernández P, Gómez Otero J, Goñi R (eds) Desde el País de los Gigantes. Perspectivas arqueológicas en Patagonia, tomo II. Universidad Nacional de la Patagonia Austral, Río Gallegos, pp 455–469

Piana EL, Orquera LA (2007) Diferencias regionales y temporales en el litoral sudoccidental de Sudamérica. In: Morello F, Martinic M, Prieto A, Bahamonde G (eds) Arqueología de Fuego-Patagonia. Levantando piedras, desenterrando huesos y develando arcanos. CEQUA, Punta Arenas, pp 311–323

Quiroz D, Carreño G, de la Fuente P (2016) El procesamiento de ballenas varadas en las costas de la Patagonia occidental (a propósito de un grabado de mediados del siglo XIX). Magallania 44(1):57–72. https://doi.org/10.4067/S0718-22442016000100004

Rabassa J, Heusser C, Rutter N (1990) Late Glacial and Holocene sea transgression in the Beagle Channel (Tierra del Fuego). Quat S Am Antarct Penins 7:335–360

Reyes O (2021) The settlement of the Chonos Archipelago, Western Patagonia. Springer, Cham, Chile

San Roman M, Björck S, Nilsson B, Sjöström A, Borrero LA, Trejo V, Martínez I, McCulloch R, Huidobro C, Torres J, Mardones J, Lukoviek E, Morello F (2023) Early maritime peopling of the Pacific Tundra and the first archaeological records from Brecknock Peninsula, Tierra del Fuego (54° S/72° W). In: Abstracts of the INQUA, Roma, 14–20 July 2023

Sierralta Navarro SG (2024) El ritmo de las piedras. Economía de la tecnología lítica en los archipiélagos occidentales de Patagonia. Un estudio desde Reloncaví y el norte de Chiloé. Ph.D. dissertation, Universidad Nacional del Centro de la Provincia de Buenos Aires

Tafuri MA, Zangrando AF, Tessone A, Kochi S, Moggi Cecchi J, Di Vincenzo F, Profico A, Manzi G (2017) Dietary resilience among hunter-gatherers of Tierra del Fuego: isotopic evidence in a diachronic perspective. PLoS One 12(4):e0175594. https://doi.org/10.1371/journal.pone.0175594

Tivoli AM, Zangrando AF (2011) Subsistence variations and landscape use among maritime hunter-gatherers. A zooarchaeological analysis from the Beagle Channel (Tierra del Fuego, Argentina). J Archaeol Sci 38:1148–1156. https://doi.org/10.1016/j.jas.2010.12.018

Vázquez MM (2024) Movilidad y Subsistencia de los cazadores-recolectores de Península Mitre, extremo sudeste de Tierra del Fuego. Ph.D. dissertation, Universidad de Buenos Aires

Vita-Finzi C, Higgs ES, Sturdy D, Harris J, Legge AJ, Tippett H (1970) Prehistoric economy in the Mount Carmel area of Palestine: site catchment analysis. Proc Prehist Soc 36:1–37. https://doi.org/10.1017/S0079497X00013074

Weddell J (2006 [1825]) Un viaje hacia el Polo Sur realizado en los años 1822–1824. Eudeba, Buenos Aires

Weisler MI (1995) Henderson island prehistory: colonization and extinction on a remote Polynesian island. Biol J Linn Soc 56:377–404. https://doi.org/10.1111/j.1095-8312.1995.tb01099.x

Zangrando AF (2009) Historia evolutiva y subsistencia de cazadores-recolectores marítimos de Tierra del Fuego. Sociedad Argentina de Antropología, Buenos Aires

Zangrando AF, Borrazzo K, Tivoli AM, Alunni DV, Martinoli MP (2014) El sitio Heshkaia 35: nuevos datos sobre la arqueología de Moat (Tierra del Fuego, Argentina). Rev Mus Antropol 7:11–24. https://doi.org/10.31048/1852.4826.v7.n1.9090

Zangrando AF, Panarello H, Piana EL (2014b) Zooarchaeological and stable isotopic assessments on pinniped-human relations in the Beagle Channel (Tierra del Fuego, Southern South America). Int J Osteoarchaeol 24(2):231–244. https://doi.org/10.1002/oa.2352

Zangrando AF, Tessone A, Vázquez M (2009) El uso de espacios marginales en el Archipiélago Fueguino: implicaciones de la evidencia arqueológica de Bahía Valentín. In: Salemme M, Santiago F, Álvarez M, Piana E, Vázquez M, Mansur E (eds) Arqueología de Patagonia: una mirada desde el último confín. Utopías, Ushuaia, pp 47–62

Zangrando AF, Tivoli AM, Fernández Ropero C, Pinto Vargas G (2021) Contextos arqueológicos en paleocostas: un nuevo caso de estudio en Bahía Moat (Tierra del Fuego). Paper presented at the XI Jornadas de Arqueología de la Patagonia 2021, Universidad Austral de Chile, Puerto Montt, 29 Nov–3 Dec 2021

Zangrando, AF, Martinoli MP, Tessone A, Vázquez MM, Alunni DV, Franch Bach A, Hernández Herrero O, Fernández Ropero MC, Pinto Vargas G, Tivoli AM (2024) El uso humano de espacios externos en el archipiélago de Tierra del Fuego durante el Holoceno: el caso de Isla de los Estados. Bol Soc Chil Arqueol 57. https://doi.org/10.56575/BSCHA.05700240843

Zangrando AF, Vázquez M, Tessone A (eds) (2011) Los cazadores-recolectores del extremo oriental fueguino. Arqueología de Península Mitre e Isla de los Estados. Sociedad Argentina de Antropología, Buenos Aires

MIX
Papier aus verantwortungsvollen Quellen
Paper from responsible sources
FSC® C105338
www.fsc.org

If you have any concerns about our products,
you can contact us on
ProductSafety@springernature.com

In case Publisher is established outside the EU,
the EU authorized representative is:
Springer Nature Customer Service Center GmbH
Europaplatz 3, 69115 Heidelberg, Germany

Printed by Libri Plureos GmbH
in Hamburg, Germany